the Heavens

A DIFFERENT VIEW

Danny Faulkner

General Editor

First printing: November 2021

Master Books®, P.O. Box 726, Green Forest, AR 72638

Master Books® is a division of
New Leaf Publishing Group, Inc.

ISBN: 978-1-68344-282-0
ISBN: 978-1-61458-792-7 (digital)
Library of Congress Number: 2021946695

Designed by Left Coast Design

Scripture quotations are from the ESV® Bible (The Holy Bible, English Standard Version®), copyright © 2001 by Crossway, a publishing ministry of Good News Publishers. Used by permission. All rights reserved.

Please consider requesting that a copy of this volume be purchased by your local library system.

Printed in the United States of America

Please visit our website for other great titles:
www.masterbooks.com

For information regarding author interviews, please contact the publicity department at (870) 438-5288.

ACKNOWLEDGEMENTS

This book would not be possible without the kind contributions of photographs from several people. While I retain copyright of the text and my photographs, the creators of the respective photographs retain the copyrights to their work.

NGC 1333 (Glen Fountain) ▲

DEDICATION

To Tom Vail, who, by publishing *Grand Canyon: A Different View,* inspired me to write this book. Furthermore, by making me part of Canyon Ministry's raft trips in the Grand Canyon, Tom made it possible for me to take some of these photos. Thank you very much, my friend.

CONTENTS

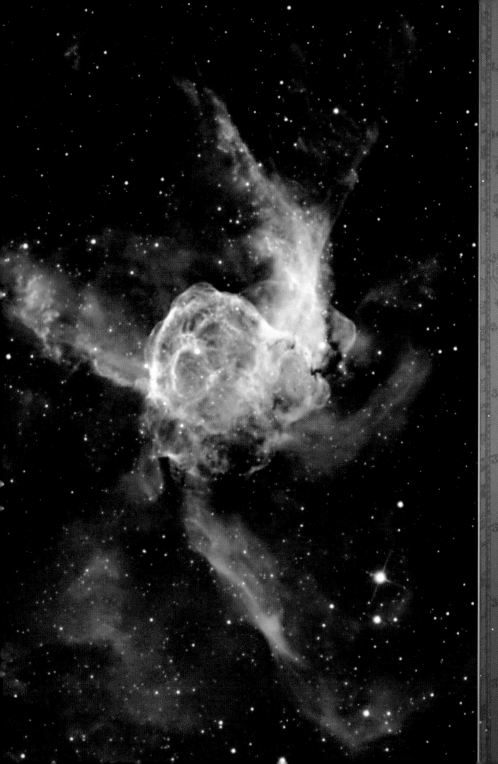

I think that astronomy is a special science, a sort of bridging of art and science. Of course, as an astronomer, I may be biased.

The Bible singles out things astronomical to underscore that there must be a Creator.

THE HEAVENS DECLARE THE GLORY OF GOD,
AND THE SKY ABOVE PROCLAIMS HIS HANDIWORK.

Psalm 19:1

WHEN I LOOK AT YOUR HEAVENS,
THE WORK OF YOUR FINGERS,
THE MOON AND THE STARS,
WHICH YOU HAVE SET IN PLACE,
WHAT IS MAN THAT YOU ARE MINDFUL OF HIM,
AND THE SON OF MAN THAT YOU CARE FOR HIM?

Psalm 8:3–4

There is something about the human soul that is in wonder of the heavens above. This contemplation ought to lead us to consider our Creator. More importantly, it should prompt us to ask the big questions about life — who we are, where we came from, and where we are going.

Thor's Helmet, NGC 2359 (Glen Fountain) ◄
Horsehead Nebula (Glen Fountain) ►

introduction

While this general revelation from the world around us can convince us that there is a God, it is inadequate to tell us all that we need to know. For all we need to know about God, we must turn to His special revelation, the Bible.

FOR WHAT CAN BE KNOWN ABOUT GOD IS PLAIN TO THEM, BECAUSE GOD HAS SHOWN IT TO THEM. FOR HIS INVISIBLE ATTRIBUTES, NAMELY, HIS ETERNAL POWER AND DIVINE NATURE, HAVE BEEN CLEARLY PERCEIVED, EVER SINCE THE CREATION OF THE WORLD, IN THE THINGS THAT HAVE BEEN MADE. SO THEY ARE WITHOUT EXCUSE.

Romans 1:19–20

Astrophotography (photography of astronomical bodies) has come a long way. Most astronomical objects are very faint, requiring long exposure times to capture them in photographs. For many years, astrophotos were taken with photographic emulsions (film). Modern electronic cameras are at least 50 times more sensitive than old emulsions. What used to take a minute exposure now can be done in a second. A minute exposure now captures what took an hour in the past. The quality and size of telescopes available to amateurs have improved tremendously. And then there is Adobe Photoshop and other electronic aids and techniques that can greatly enhance astrophotos. Consequently, amateur astronomers now can take photographs that only professional astronomers could do a few decades ago. There is a downside to this — the eye can collect light for only a tiny fraction of a second. When looking through a telescope, many people expect galaxies and other faint objects to look exactly as what they see in photographs. As a result, most people aren't very

impressed when seeing a faint blob of light through a
telescope. It helps to explain that they might be looking
at the combined light of hundreds of billions of stars in
a galaxy that is many millions of light years away.

In this book I share some of my astrophotos, along
with many others that a few amateur astronomers have
taken. It is our pleasure to share with you the beauty
and wonder of the creation as captured by our work.
But, more importantly, it is our desire that these photos
draw you closer to our Creator.

Everyone knows that each day the sun rises in the east, moves across the sky, and sets in the west. And most people know that this is the result of the earth's rotation on its axis. When the sun sets, the earth keeps rotating. If it didn't, then the sun

Star Trails

wouldn't rise tomorrow morning. Less well known is the fact that the moon and most stars also rise in the east and set in the west. However, toward the north, there are some stars that neither rise nor set. Instead, these stars perpetually move in circles around a spot in the northern part of the sky, a spot we call the north celestial pole. Indeed, all celestial objects exhibit this daily circular motion, though most, due to their great distance from the north celestial pole, have their circles interrupted by setting. We say that the stars that are always up are circumpolar, meaning "around the pole."

The star very close to the center of the circular motion in these photos is Polaris. Contrary to popular misconception, Polaris isn't the brightest star in the sky, nor is it the first star visible at night. Rather, its unique status is that it is so close to the north celestial pole that, to the unaided eye, Polaris appears motionless in the north direction. It is the only star that appears in the same spot all the time. This property allows us to find the north direction at night and is why we often call Polaris the North Star. Because the earth is spherical, the angle that Polaris makes with the horizon is roughly equal to one's latitude.

The north celestial pole isn't visible from the earth's Southern Hemisphere. Instead of spinning around the north celestial pole in the northern part of the sky, stars in the Southern Hemisphere appear to spin around the south celestial pole in the southern part of the sky. But, unlike in the Northern Hemisphere, where the stars appear to spin counterclockwise, stars in the Southern Hemisphere appear to spin clockwise. Furthermore, there is no bright star close to the south celestial pole. Therefore, there is no "south star."

▲ *Star trails over the Blue Ridge Mountains of North Carolina* (Danny Faulkner)
▼ *Star trails* (Jim Bonser)
▶ *Star trails over Grand View Camp, Eagar, Arizona* (Danny Faulkner)

And God said, "Let there be lights in the expanse of the heavens to separate the day from the night. And let them be for signs and for seasons, and for days and years, and let them be lights in the expanse of the heavens to give light upon the earth." And it was so. And God made the two great lights — the greater light to rule the day and the lesser light to rule the night — and the stars.

Genesis 1:14–16

Star trails over the Creation Museum (Deb Bonser) ▲
Star trails over the Ark Encounter (Jim Bonser) ▶

The Milky
Way is a
faint band of light
extending across the
sky. Most people never fully
appreciate the Milky Way because
city lights usually overwhelm the
Milky Way's dim glow. Even the light of
the moon hampers the view of the Milky Way.
People in the past probably enjoyed the Milky Way
far more than we do today. With no artificial lights,
the Milky Way would have been prominent on many

the Milky Way

clear, moonless nights. Furthermore, without electricity, there
was no TV or lights to keep people occupied inside at night.
When not sleeping, people probably were driven outside on
warm summer evenings, when the Milky Way is most prominent,
at least in the Northern Hemisphere. I enjoy taking photographs of the

Milky Way in late summer and early autumn, when its most intense part seems to make the best "pose" with objects on the horizon as it sets. Notice the dark lanes in the Milky Way. These are caused by large clouds of dust blocking the view of more distant stars.

The Milky Way is our galaxy, a vast collection of billions of stars held together by gravity. The galaxy is round and flat with a bulge in its center, sort of like an over-easy egg. Coming off the nuclear bulge are graceful spiral arms. The spiral arms contain much dust and gas, along with many bright stars. The sun is located outside of a spiral arm, about halfway from the center to the edge. The diameter of the Milky Way is about 100,000 light years, so the sun is about 25,000 light years from the center. On summer and early autumn evenings in the Northern Hemisphere, the central bulge of the Milky Way is visible.

▶ *Summer Milky Way showing nuclear bulge near trees* (Danny Faulkner)

◀ *View of the summer Milky Way from the Grand Canyon. The bright star is Jupiter.* (Danny Faulkner)

Astronomers estimate that there are a couple hundred billion stars in the Milky Way Galaxy. There are many other galaxies — the current best estimate is that there are at least two hundred billion galaxies similar in size to our own Milky Way. If each one of those galaxies contains the same number of stars as the Milky Way, how many stars are there?

I'll leave the answer to that question as an exercise for the reader. Just keep in mind that these are just estimates. But, according to Psalm 147:4 and Isaiah 40:26, God counts the number of the stars, and He names all of them!

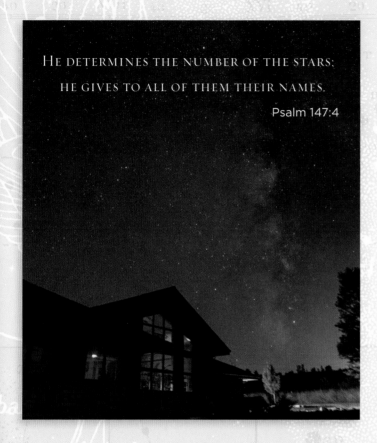

He determines the number of the stars;

he gives to all of them their names.

Psalm 147:4

Lift up your eyes on high and see:

who created these?

He who brings out their host by

number, calling them all by name;

by the greatness of his might

and because he is strong in power,

not one is missing.

Isaiah 40:26

▲ *Milky Way* (Jim Bonser)
▶ *Milky Way* (Jim Bonser)
◀ *Center of the Milky Way with a few clouds* (Danny Faulkner)

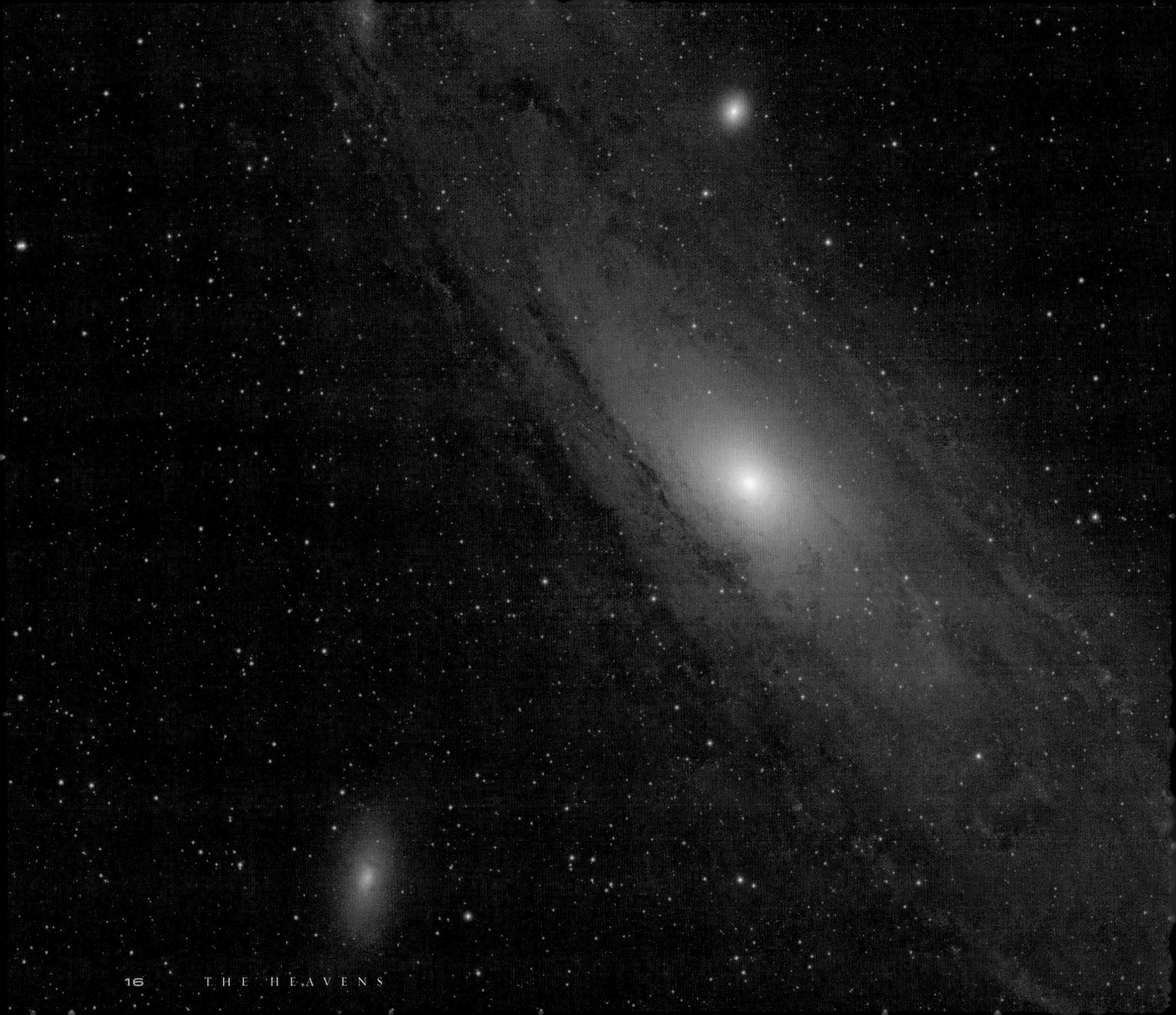

The Andromeda Galaxy (M31, or NGC 224) is the closest galaxy of any size. On a clear, dark autumn evening, it is easy to spot M31 in the constellation Andromeda. About 2½ million light years away, the Andromeda Galaxy is thought to be slightly larger than the Milky Way.

The Andromeda Galaxy also contains more stars than our Milky Way Galaxy does. In either photo, notice the two satellite galaxies — M32 (NGC 221) above the center of M31, and M110 (NGC 205) to the lower left of the center of M31. Smaller satellites of larger galaxies are common.

▲ *M31* (Jim Bonser)
◀ *M31* (Glen Fountain)

The two largest satellite galaxies of the Milky Way. The Large Magellanic Cloud (LMC) is the nebulosity just to the right of center. The Small Magellanic Cloud (SMC) is the smaller nebulosity below and slightly to the right of the LMC, near the trees. The Milky Way is to the left. (Danny Faulkner)

For instance, the Milky Way has two satellite galaxies, the Large Magellanic Cloud (LMC) and the Small Magellanic Cloud (SMC) (NGC 292). However, the LMC and SMC are near the south celestial pole, so they aren't visible from much of the Northern Hemisphere.

M31 is the largest galaxy in the Local Group, a small collection of galaxies consisting of M31, the Milky Way, M33, and about 50 dwarf galaxies. M33 (NGC 598) is sometimes called the Triangulum Galaxy because of its location in the direction of the constellation Triangulum. M33 is thought to be a little farther away than M31, perhaps as much as three million light years. M33 is quite a bit smaller, fainter, and contains fewer stars than the two other large galaxies in the Local Group.

▶ *M33* (Glen Fountain)

A bit farther away is M81 (NGC 3031). M81 is about 12 million light years away in the direction of the constellation Ursa Major. M81 is the most distant object that normally can be seen with the naked eye. However, seeing it is very difficult — it requires very clear, dark skies and total dark adaptation. I once saw it from 7,000 feet in Arizona after dark adapting my eyes for four hours. Astronomers think that the core of M81 harbors a black hole that is 70 million times more massive than the sun (in astronomy, "massive" refers to how much matter something contains, not its physical size, such as diameter). M81 is the dominant galaxy of the M81 group, consisting of M81 and a few score smaller galaxies.

Much farther out is NGC 2903. It is about 30 million light years away in the direction of the constellation Leo. All the large galaxies that we've seen thus far are spiral galaxies — flat, round distributions of stars with central bulges from which spiral arms gracefully flow outward. However, NGC 2903 sports a modest bar across its nucleus, from which spiral arms are attached. In recent years, astronomers have come to realize that our own Milky Way Galaxy has a small bar across its nucleus as well. However, the dust between us and the galactic center dust prevents us from seeing this bar.

There are other galaxies in Leo, such as the Leo Triplet. The galaxies are, clockwise from top right, M65 (NGC 3623), M66 (NGC 3627), and NGC 3628. Both M65 and M66 are barred spirals, but NGC 3628 is a normal spiral galaxy. The Leo Triplet is about 35 million light years away.

M96 (NGC 3368), a spiral galaxy about 35 million light years away, also is in Leo. You may be wondering what these designations mean, the letter M or the letters NGC followed by numbers. The M stands for the Messier catalogue. Charles Messier was a French astronomer who discovered 13 comets in the late 18th century. As Messier searched for comets, he often saw star clusters, galaxies, and nebulae (clouds of gas in space) that superficially resembled comets. To avoid confusion (and unwarranted excitement over supposedly discovering another "comet"), Messier assembled a catalogue of a little more than a hundred objects in no order. The German-born English astronomer William Herschel was a contemporary of

Leo Triplet (Jim Bonser) ▲
M96 (Jim Bonser) ▶

Messier. Armed with a larger telescope than Messier had, Herschel embarked on a more systematic search for star clusters, galaxies, and nebulae, eventually discovering more than 2,400 objects. Herschel published three versions of his *Catalogue of Nebulae and Clusters of Stars*. William Herschel's son John Herschel expanded upon his father's work, eventually publishing his *General Catalogue of Nebulae and Clusters* in 1864. Later, the Danish-born British astronomer John Dreyer further expanded the work of both Herschels and published it as the *New General Catalogue of Nebulae and Clusters of Stars* in 1888. The NGC has 7,840 entries in order of west to east in the sky. It's common for astronomical objects to have multiple catalogue designations even beyond these two catalogues. The usual practice is to use the Messier number if an object is in that catalogue and to use the NGC number if it is not a Messier object.

M51 (NGC 5194) frequently is called the Whirlpool Galaxy because of its grand design spiral structure. It clearly is interacting with a smaller companion galaxy. M51 is 23 million light years away in the direction of the constellation Canes Venatici.

The galaxy NGC 891 is about 37 million light years away in the constellation Andromeda. This is a spiral galaxy viewed edge-on. The nuclear bulge clearly is visible, as well as the blocking of light by dust along the plane of the galaxy. These resemble similar features we see in the Milky Way. Because NGC 891 is about the same size and brightness of the Milky Way, it may tell us what our own galaxy might look like if viewed edge-on.

M51 in black and white (Glen Fountain) ▲
M51 in color (Glen Fountain) ◄
NGC 891 (Jim Bonser) ►

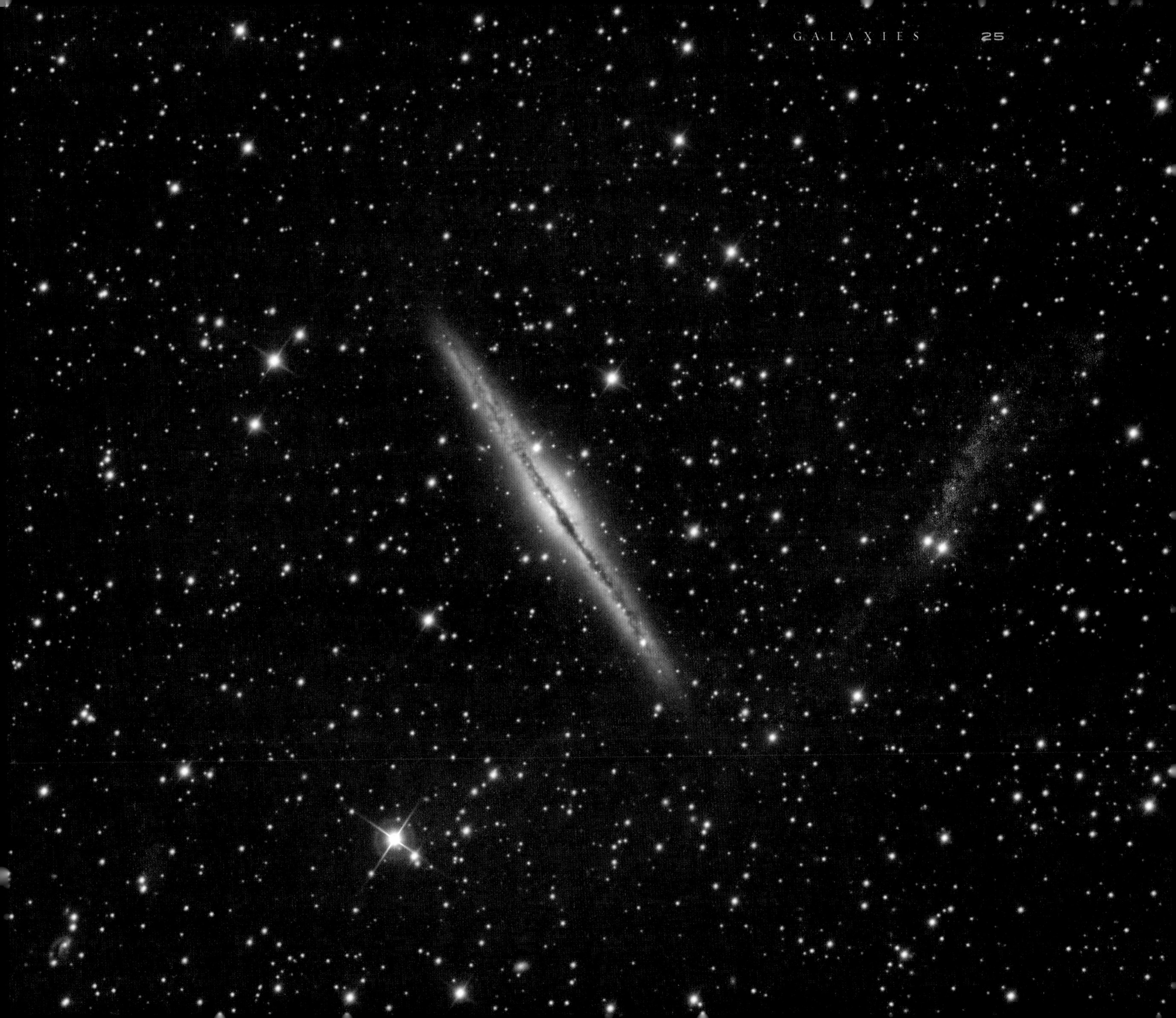

Another edge-on galaxy is NGC 4565, in the constellation Coma Berenices. Because of its narrow profile, NGC sometimes is called the Needle Galaxy. NGC 4565 is a very large, bright galaxy, probably larger than the Andromeda Galaxy. Its distance is unsure, somewhere between 30 and 50 million light years away. Notice NGC 4562, the faint spiral galaxy at the top of the photograph. This galaxy is roughly the same distance as NGC 4565.

Galaxies tend to clump together into clusters of galaxies. The Hercules Cluster is a cluster of hundreds of galaxies about 500 million light years away. Its brightest member is the giant elliptical galaxy NGC 6041, to the upper left in the photo on page 28. The Hercules Cluster sometimes is called Abell 2151, from the *Abell Catalog of Rich Clusters of Galaxies*, compiled by the astronomer George Abell and some of his collaborators.

A bit closer to home is the Virgo Cluster, a collection of about 1,500 galaxies a little more than 50 million light years away. Markarian's Chain is a part of the Virgo Cluster.

▲ *Markarian's Chain* (Jim Bonser)
◀ *NGC 4565* (Jim Bonser)

THE LIGHT TRAVEL TIME PROBLEM

The galaxies illustrated here are millions of light years away. Most galaxies are much farther. The most distant galaxies observed are several billion light years away. This raises an important question: How could the light from these distant objects have reached us if the creation is only thousands of years old, as the Bible indicates? Creationists call this the light travel time problem. There have been several suggested solutions to the light travel time problem, but there is not one generally accepted explanation.

We need to keep in mind that everything about the creation week was miraculous. The most straightforward answer may be that God simply performed a miracle to bring the light to earth on day four, the same day that He made the astronomical bodies. Creation of the universe required the introduction of space, time, matter, and energy. Compared to that, getting the light from the most distant parts of the universe is rather trivial.

Hercules Cluster, Abell 2151 (Glen Fountain) ▲
Hercules Cluster, Abell 2151, labeled Annotated (Glen Fountain) ▶

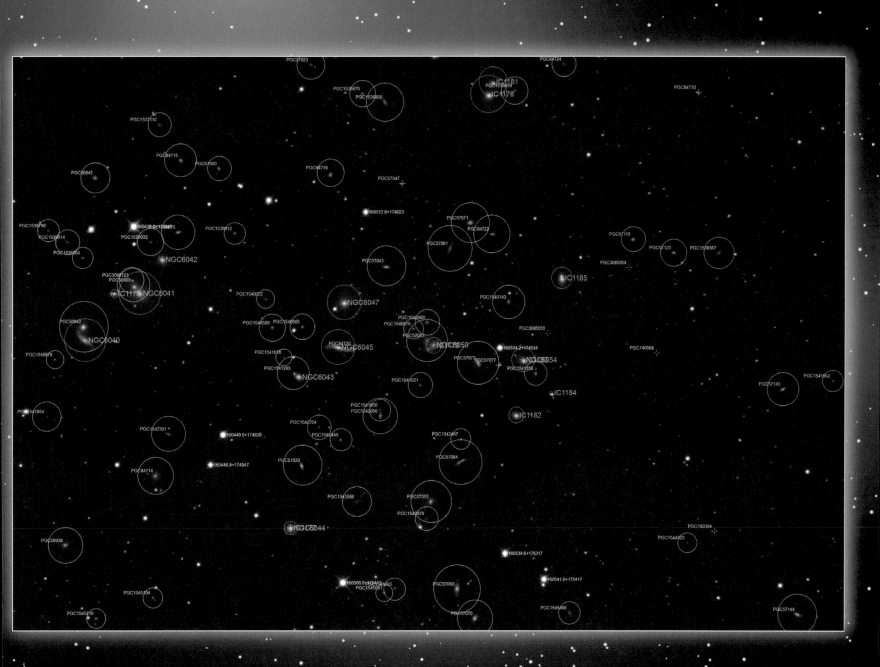

Orion, the Pleiades, & OTHER GROUPS OF STARS

Many bright stars naturally seem to group together into patterns that we call constellations. There are many different constellation systems around the world. However, the dominant set of constellations today comes from Western culture. We get our Western constellations from the ancient Greeks, who in turn borrowed from even earlier civilizations, probably in the ancient Near East. In the early 2nd century, the Greek astronomer Claudius Ptolemy catalogued more than a thousand stars into 48 constellations, but there is evidence that these constellations are more ancient still. Most of Ptolemy's 48 constellations survive today, but we've added more in recent centuries. Today, astronomers recognize 88 constellations. These constellations have defined boundaries so that the 88 constellations completely fill the sky. Therefore, there won't be any more constellations.

One of the most popular constellations is Orion, the hunter. The most prominent feature of Orion is the three stars in a line that comprise his belt. Astronomers think that these three stars are between one and two thousand light years away. This makes these stars some of the most distant among the more obvious stars in the sky. Orion is one of two constellations that are mentioned in the Bible.

He who made the Pleiades and Orion,
and turns deep darkness into the
morning and darkens the day into night,
who calls for the waters of the sea and
pours them out on the surface of the
earth, the Lord is his name. . . .

Amos 5:8

▲ *The Pleaides* (Jim Bonser)
▶ *Orion through clouds* (Jim Bonser)

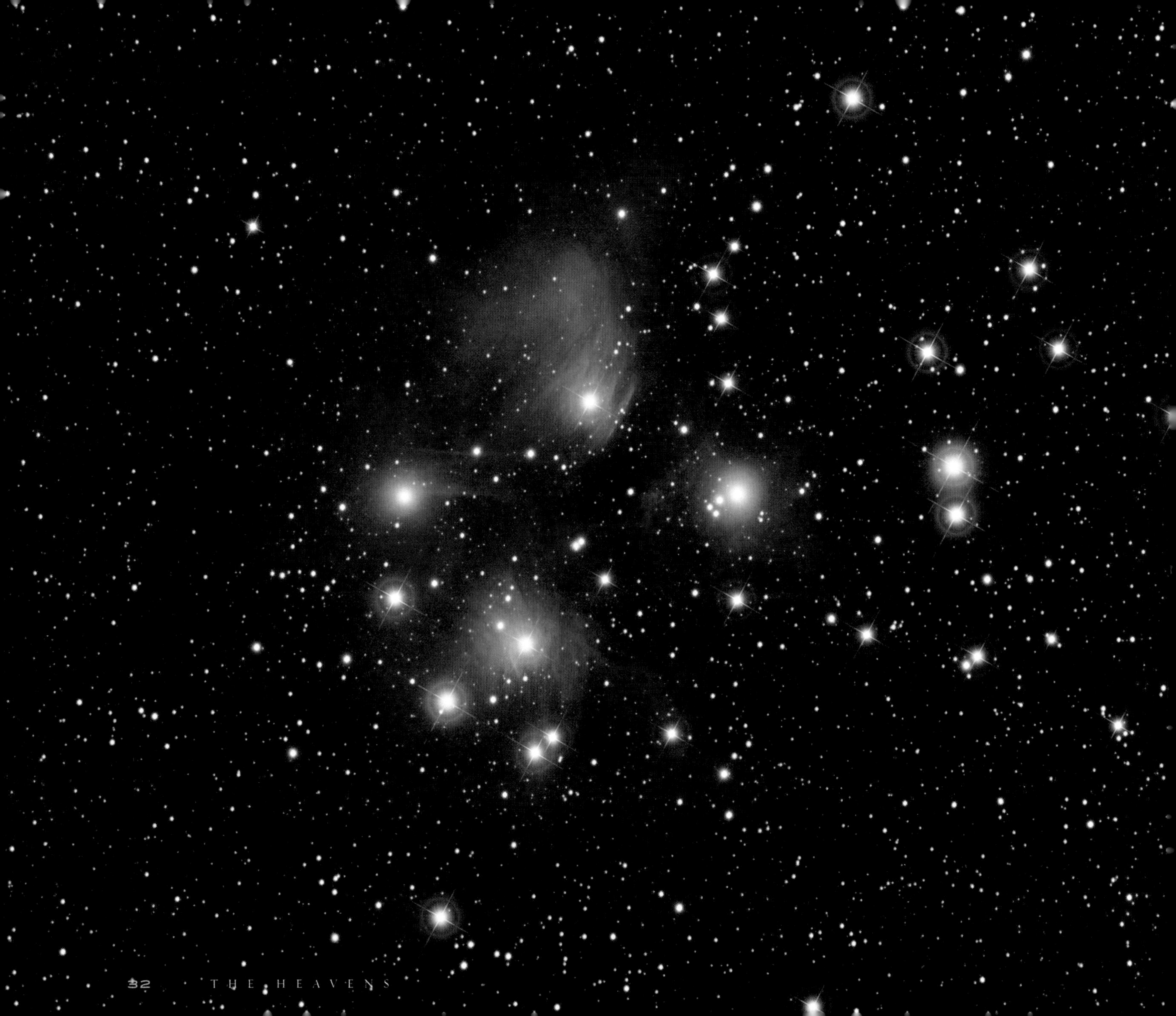

Each of the three times Orion is mentioned in the Bible, the Pleiades are included. The Pleiades is a star cluster not too far away from Orion in our sky, which may explain why the two always are mentioned together in Scripture. A star cluster is a gravitationally bound group of stars containing hundreds, or even thousands, of stars. The Pleiades contains at least a thousand stars. However, most of those stars are faint. Only about a half-dozen stars are readily visible to the naked eye on a clear, dark night. This is why the Pleiades also is called the Seven Sisters. The Japanese call the Pleiades Subaru (you may recognize the pattern of the six brightest members of the Pleiades in the Subaru symbol).

▲ *Close-up of Orion's belt and sword, showing the Orion Nebula in the sword* (Jim Bonser)
▶ *Cropped and enlarged view of the above* (Jim Bonser)
◀ *The Pleiades* (Glen Fountain)

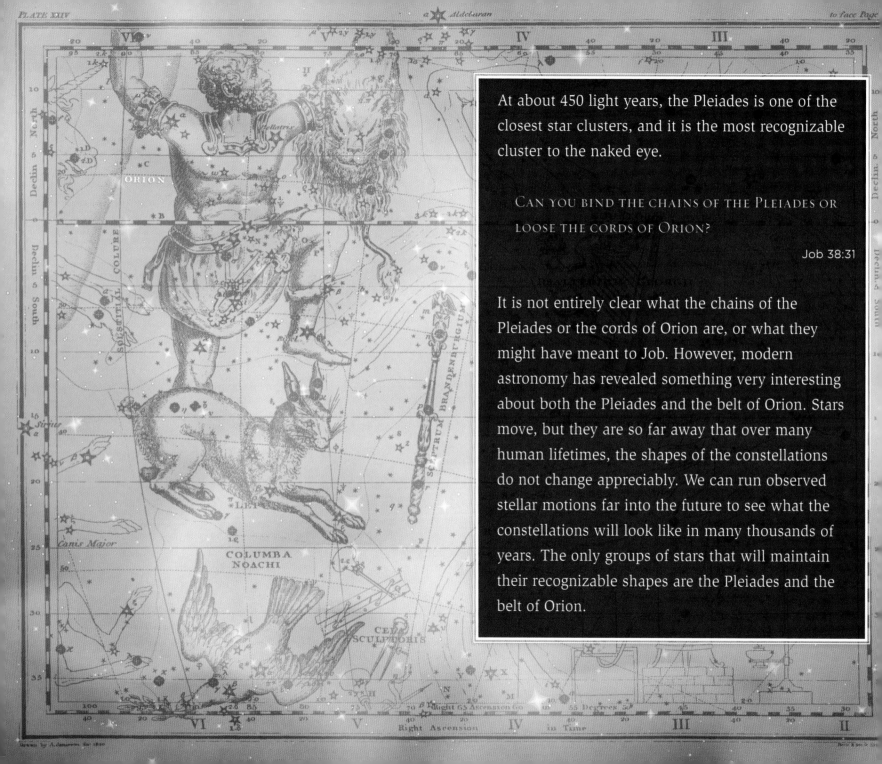

PLATE XXIV   α ✶ Aldebaran   to face Page

At about 450 light years, the Pleiades is one of the closest star clusters, and it is the most recognizable cluster to the naked eye.

CAN YOU BIND THE CHAINS OF THE PLEIADES OR LOOSE THE CORDS OF ORION?

Job 38:31

It is not entirely clear what the chains of the Pleiades or the cords of Orion are, or what they might have meant to Job. However, modern astronomy has revealed something very interesting about both the Pleiades and the belt of Orion. Stars move, but they are so far away that over many human lifetimes, the shapes of the constellations do not change appreciably. We can run observed stellar motions far into the future to see what the constellations will look like in many thousands of years. The only groups of stars that will maintain their recognizable shapes are the Pleiades and the belt of Orion.

. . . WHO MADE THE BEAR AND ORION, THE
PLEIADES AND THE CHAMBERS OF THE SOUTH. . . .

Job 9:9

Of course, the Bear mentioned here is the constellation
Ursa Major, or the Big Bear. Both the Big Bear and
the Little Bear are close to the north celestial pole,
so they are sort of sentinels of that part of the sky as
they spin around it each day. The Big Bear is a large
constellation. However, seven of the more prominent
stars of the Big Bear have a familiar shape — the Big
Dipper in the United States, but the Plough in the
British Isles. The "chambers of the south" probably
refers to other constellations in general.

▲ *The Big Dipper is right above the trees* (Jim Bonser)

Technically, Orion appears a fourth time in the Old Testament:

> FOR THE STARS OF THE HEAVENS AND THEIR
>   CONSTELLATIONS WILL NOT GIVE THEIR
> LIGHT;
> THE SUN WILL BE DARK AT ITS RISING,
> AND THE MOON WILL NOT SHED ITS LIGHT.
>
> Isaiah 13:10

The word translated "constellations" here is the plural form of Orion (the Hebrew word for Orion is one of the two words for "fool"). Given the context and usage of the plural, translating the word as "constellations" is proper.

IS THERE A GOSPEL IN THE STARS?

Some Christians believe that the gospel account is told in the constellations. We call this theory the gospel in the stars. According to this theory, God ordained this gospel message for man prior to God's direct revelation in the Bible. However, there are numerous problems with this theory. For instance, while Psalm 147:4 and Isaiah 40:26 tell us that God has named all the stars, there is no indication that God has shared His names of the stars with man. It is likely that our names for the stars originated with man. Orion offers a good example. Supporters of the gospel in the stars theory find meaning in our name for Orion, and they suggest that Orion represents Jesus Christ. However, God's name for Orion (as indicated by the three times that Orion appears in the Old Testament) is one of the two Hebrew words meaning "fool." Christ is no fool, and it is blasphemous to make such a suggestion.

The Pleiades is an example of an open star cluster. There are many other open star clusters. They include M36 (NGC 1960), M38 (NGC 1912), and the Owl Cluster (NGC 457). The Owl Cluster gets its name from its prominent two brightest stars that suggest an appearance of an owl (the two stars represent the owl's eyes). Open star clusters contain at most a few thousand stars and have an irregular appearance. If you spin this book upside down, you will see that each cluster looks different. The Owl Cluster is about 8,000 light years away in the constellation Cassiopeia. M36 is more than 4,000 light years away in the constellation Auriga. M38 also is in Auriga, about 4,500 light years away.

◀ *The open cluster M36* (Jim Bonser)
◀ *The open cluster M38* (Jim Bonser)
▶ *The Owl Cluster* (Jim Bonser)

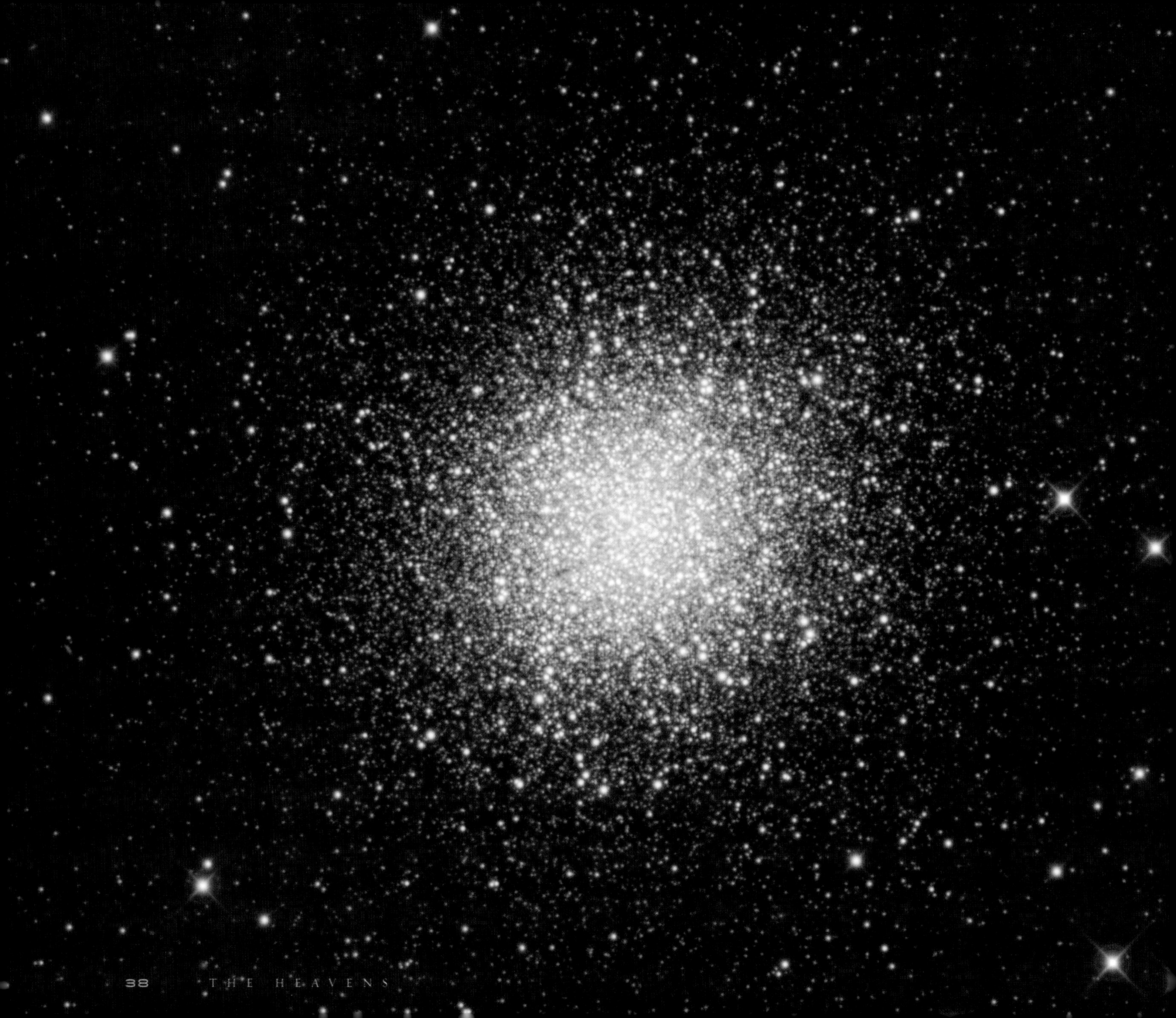

Contrast the appearance of open star clusters to globular star clusters, as with M13 (NGC 6205) and M3 (NGC 5272). Globular clusters contain far more stars — up to a few hundred thousand. Globular clusters are centrally condensed and are in a globe-like shape. This gives globular clusters spherical symmetry — if you spin this book, globular clusters don't look any different. Located in the constellation Hercules, M13 is one of the brightest and grandest globular clusters. In a very dark sky, it is visible to the naked eye as a faint fuzzy star. For a globular cluster, M13 is relatively close, only about 22,000 light years away. M3 is a bit farther away, about 34,000 light years, in the constellation Canes Venatici.

▲ *The globular cluster M3* (Glen Fountain)
◀ *The globular cluster M13* (Glen Fountain)

To the ancients, any luminous body in the sky was a star. We define things differently today, so the ancient classification of stars included some things we now don't think of as being stars, such as planets, asteroids, meteors (shooting stars), and comets (hairy stars).

Ancient astronomers also included with stars a few fuzzy patches of light in the sky. They called them *nebulae*, meaning "cloudy." With the invention of the telescope four centuries ago, astronomers naturally viewed nebulae (the singular is *nebula*) to determine their nature. Some nebulae turned out to be clusters of stars. A few other nebulae turned out to be distant galaxies. Still other nebulae turned out truly to be clouds of gas. Astronomers today retain the use of the word nebulae to refer to these clouds of gas in space.

The lower portion of Orion. The Orion Nebula is the red glow around the middle star in Orion's sword. (Jim Bonser) ◀
The central portion of the Orion Nebula (Glen Fountain) ▶

Nebulae

that the green-sensitive cones in my eyes are more sensitive than most people's. With greater sensitivity, my eyes are picking up a couple of green emission lines due to oxygen and nitrogen. The Orion Nebula is one of my favorite things to look at through a telescope. At the center

The finest example of a nebula is the Orion Nebula, also known as Messier 42 or NGC 1976. Though it is 1,300 light years away, the Orion Nebula is visible to the naked eye as a fuzzy patch around the middle star of Orion's sword. In the photos it shows up as the red glow around the middle star in the sword of Orion. Our eyes don't see color very well in dim light, so nebulae and other faint objects normally look black and white to most people. However, I sometimes see in the Orion Nebula a faint iridescent green color. I suppose

of the Orion Nebula is the Trapezium, another star cluster. The four bright stars in this cluster give it its name. In long-exposure photographs, the Trapezium is overexposed, but through a telescope, the eye can see both the Trapezium and the nebula. Astronomers believe that ultraviolet radiation from these very hot stars is responsible for lighting up the gas in the Orion Nebula. The physics of this process is well understood, and it allows astronomers to measure properties of the Orion Nebula, such as temperature and density. It is wonderful that God has made the world so understandable so that we can learn more about the creation and hopefully come to a better appreciation of the Creator.

The Orion Nebula and nebulae like it are called HII regions. The name indicates that HII regions are made primarily of ionized hydrogen gas (clouds of neutral hydrogen gas in space are called HI regions). HII regions typically appear red in color photographs because of much emission that the ionized hydrogen produces in the red part of the spectrum as electrons recombine with hydrogen ions (protons) to form neutral hydrogen.

◀ *The Orion Nebula* (Glen Fountain)
▶ *The Trapezium* (Jim Bonser)

Below the leftmost star of Orion's belt is the Horsehead Nebula. Though very pretty, the Horsehead Nebula is very faint, much fainter than the nearby Orion Nebula. The Horsehead Nebula is about 1,500 light years away. The bright red color is due to emission of hydrogen gas. The dark horsehead silhouette is the result of a cloud of very small dust particles blocking light from more distant objects. The blue glow to the lower left of the Horsehead Nebula is the reflection nebula NGC 2023. A reflection nebula is caused by light scattering off dust particles. Reflection nebulae distinctively show up as blue, while emission nebulae appear red.

The bright star to the right of the Horsehead Nebula is the leftmost star of Orion's belt △
(the photo is oriented 90 degrees from its appearance in the sky). (Glen Fountain)
Horsehead Nebula close-up (Glen Fountain) ▷

Another emission nebula is the North American Nebula (NGC 7000), named so because of its resemblance to North America. The North American Nebula is 1,600 light years away in the direction of the constellation Cygnus.

Yet another emission nebula is the Rosette Nebula, located in the constellation Monoceros. The Rosette is about 5,200 light years away. The emission of the nebula is powered by ultraviolet light from the brighter stars of the star cluster NGC 2244 located at its center.

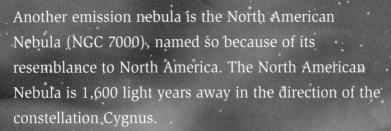

The North American Nebula (Glen Fountain) ◄
The Rosette Nebula (Glen Fountain) ▲
The Rosette Nebula (Glen Fountain) ▶

Another example of an emission nebula is the Eagle Nebula (M16, or NGC 6611), located in the constellation Serpens. In the black and white image, the namesake eagle shows up near the center. M16 is both an emission nebula and a star cluster, the brighter members of which can be seen interspersed throughout the nebula. As before, it is the hotter, brighter stars of this cluster that power the hydrogen emission with its characteristic red color. The Eagle Nebula is an estimated 7,000 light years away.

Besides emission and reflection nebula, there are other types of nebulae. One other type is planetary nebulae, of which the Dumbbell Nebula (M27, or NGC 6853) was the first discovered in 1764. Located in the constellation Vulpecula, the Dumbbell Nebula is about 1,400 light years away. The term "planetary nebulae" is a misnomer because planetary nebulae have nothing to do with planets. The name came from the fact that in small telescopes, planetary nebulae appear round, as planets do.

The Eagle Nebula. The "eagle" is near the center. (Glen Fountain) ▲
A portion of the Eagle Nebula (Glen Fountain) ◄
The Dumbbell Nebula (Glen Fountain) ▶

In this photo, you can see a faint star directly at the center of the Dumbbell. This is common in all planetary nebulae. Their central stars are very hot examples of a type of star called white dwarfs. Compared to other stars, white dwarfs are very small. However, the white dwarf at the center of the Dumbbell Nebula is much larger than most white dwarfs, about six times the size of the earth. White dwarfs are very dense. If you brought a teaspoon of this star at the center of the Dumbbell Nebula to the earth, it would weigh 50 pounds. Most white dwarfs are far denser than that.

The central stars of planetary nebulae heat the gas in the nebulae, causing the gas to glow. The central stars rapidly cool, reducing how effective the heating is. Furthermore, as planetary nebulae expand, they cool and dim. Consequently, planetary nebulae are only thousands of years old. Astronomers think that winds present in the central stars of planetary nebulae blew the gas off to form the nebulae.

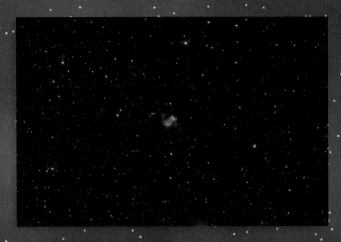

The much fainter planetary nebula M76 (NGC 650/651) is nearly twice as far away, at 2,500 light years. Because of its resemblance to the Dumbbell Nebula, M76 often is called the Little Dumbbell Nebula, or the Barbell Nebula.

The Ring Nebula (M57, NGC 6720) is another fine example of a planetary nebula. It is about 700 light years away. As with the Dumbbell Nebula, you can see the central star in the Ring Nebula. While the central stars of planetary nebulae may show up in these long-exposure photos, they are too faint to be visible when looking at these nebulae through most telescopes.

The final type of nebula is a supernova remnant. The best example of a supernova remnant is the Crab Nebula (M1, or NGC 1952). The familiar name comes from a drawing of the nebula published in 1844 by William Parsons, the Third Earl of Rosse, that resembled a crab. Located in the constellation Taurus, the Crab Nebula is about 6,500 light years away. Identification as a supernova remnant came in the 1920s, when astronomers realized that the nebula was expanding and its position coincided with a supernova that Chinese astronomers recorded in A.D. 1054. By the 1940s, astronomers realized that a star at the nebula's center had some unusual properties, which suggested that it might be responsible for the nebula. One of those unusual properties was that the star was a strong radio source.

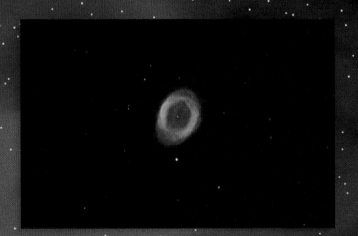

The Ring Nebula (Jim Bonser) ▲
M76 (Jim Bonser) ◄
The Crab Nebula (Glen Fountain) ▶

Things rapidly began to change in 1968, when it was learned that the central star's radio emission pulsed on and off 30 times per second. This was one of the first pulsars discovered. Pulsars don't actually pulse. Instead, astronomers think that pulsars are very small, dense, rapidly spinning objects called neutron stars. Neutron stars emit a beam of radiation, and as the stars spin, we may periodically see their beams flash on and off, much as we see the beams of a rotating search light. The discovery of pulsars was the beginning of supernova science and the detailed study of their remnants. Astronomers now believe that supernovae usually leave behind a neutron star (sometimes observed as pulsars) or black holes. There now are thousands of known pulsars and many other supernova remnants. Like planetary nebulae, supernova remnants rapidly expand and dissipate, so they don't last very long. Neutron stars are incredibly dense — a teaspoon of a neutron star brought to earth would weigh a billion tons!

the Sun

The sun is the largest body in our solar system. The sun's diameter is 109 times that of the earth's diameter. In this photograph, an image of the earth is inserted for scale. If one compares volume, the difference in size is even more impressive — you could fit more than a million earths inside the sun. The sun has 333,000 times the mass of the earth. The dark spots that you see are sunspots. As you can see, some sunspots are larger than the earth.

Notice that the sun's limb (edge) isn't as bright as its center. Astronomers call this limb darkening. Limb darkening is caused by the sun's atmosphere, the outermost region of the sun where its density decreases rapidly toward the vacuum of space. Near the center of the sun's image, its light passes perpendicular to the sun's surface to us. However, near the sun's limb, light must pass at a low angle to the surface, greatly extending the distance that the light must pass through the solar atmosphere. This blocks some of the light, causing us to see to less depth near the sun's limb than we do at the center of the sun's image. Since the temperature is cooler higher in the sun's atmosphere, the light we receive from the higher layer near the solar limb is dimmer.

The sun and sunspots with the earth to scale for comparison (Danny Faulkner) ▶

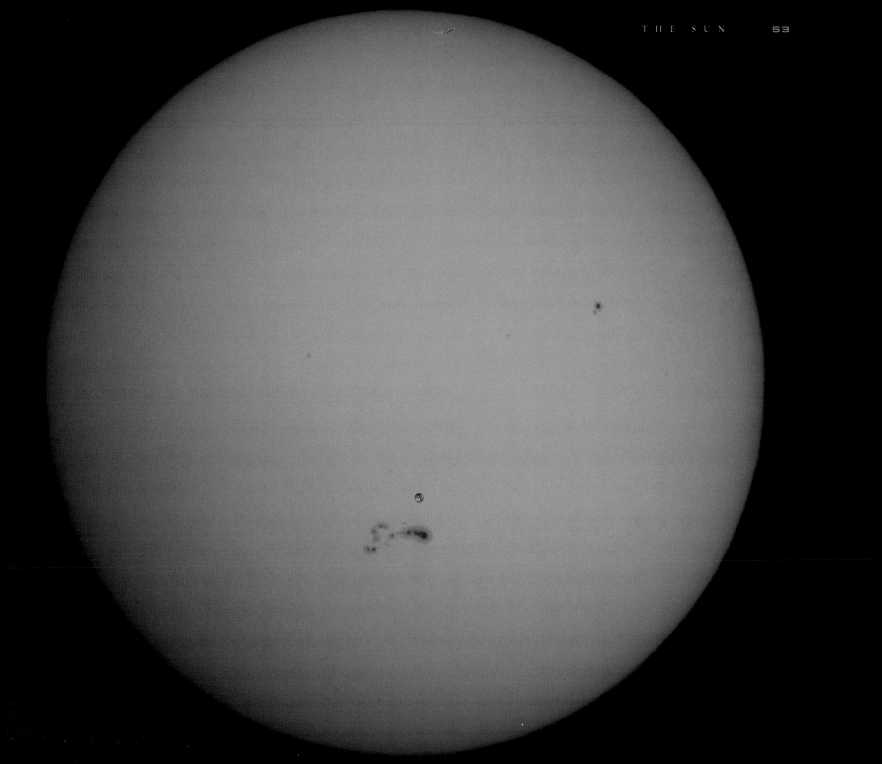

Notice the structure of sunspots. They tend to be darkest in their middles and less dark near their edges. Astronomers call these the umbra and penumbra of sunspots, respectively (we borrowed these terms from two types of eclipse shadows). Sunspots aren't truly dark at all — the light of a sunspot is bright enough to blind you. They just look dark compared to the much brighter normal parts of the photosphere, the sun's "surface." By the way, looking at the sun can be very harmful to your eyes. You should never look at the sun without the proper equipment. And even if you have the proper equipment, if you don't know how to use the equipment, you should not look at the sun. The photos that you see here were made with the proper equipment.

Sunspots last a few days to a few weeks. You can see that sunspots change over several days. One change is their structure as they appear, develop, fade, and then disappear. You also can see sunspots move across the face of the sun as the sun rotates. It takes about a month to complete one rotation. Sunspots are regions of strong local magnetic fields on the sun. The number of sunspots varies over a cycle that averages 11.2 years. Sunspot minimum occurred in 2020, shortly before this book went to press. The next sunspot maximum is expected around 2025.

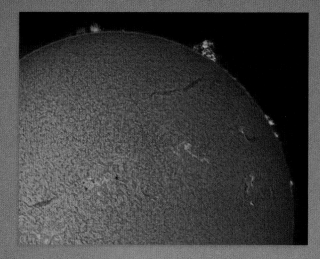

One of the more popular filters for observing the sun is H-alpha. This is the same wavelength that is responsible for much of the light from emission nebulae. It is in the red part of the spectrum, so the color of the images here are not quite correct. The sun's photosphere does not emit much light in H-alpha, but the chromosphere, a thin layer just above the photosphere, does. Therefore, in H-alpha we see the chromosphere. Above the chromosphere are prominences, the flame-like things projecting out into space beyond the sun's limb.

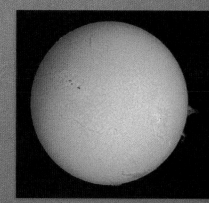

◀ *September 4, 2017 — two large groups of sunspots* (Danny Faulkner)
▶ *The sun in H-alpha light* (Jim Bonser)
▶ *The sun in H-alpha light* (Jim Bonser)

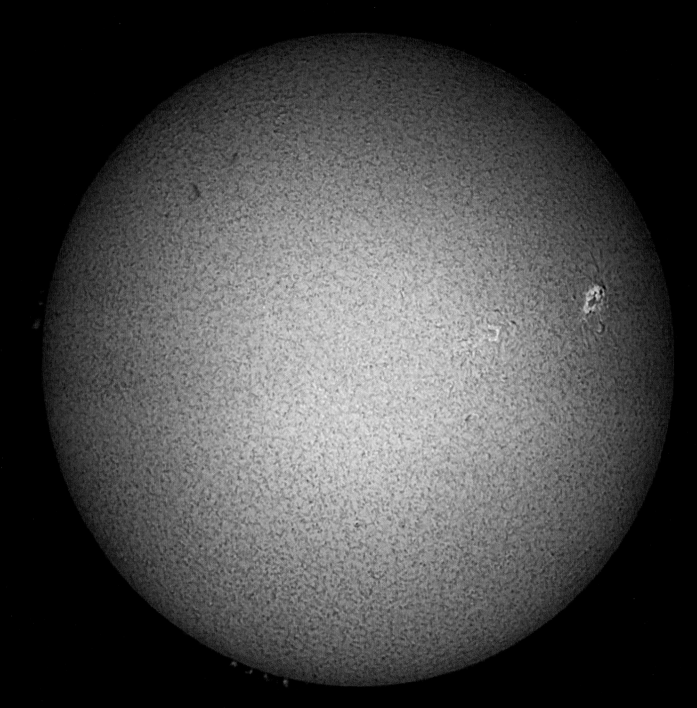

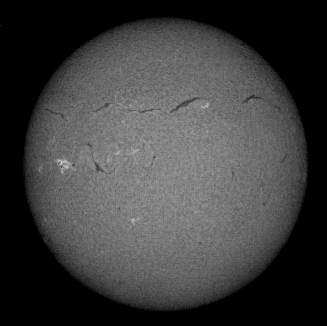

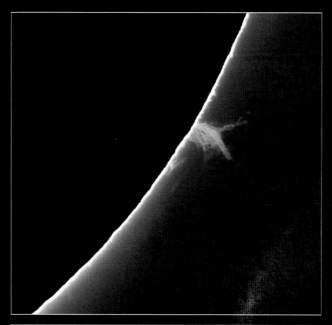

When prominences are visible against the chromosphere, they are called filaments. Filaments are the darker, rope-like features on the face of the sun in some of these photographs. Without special filters as used here, the eye cannot see prominences and filaments.

Prominences look like violent eruptions on the sun, but they aren't. Some people mistake prominences for solar flares, but solar flares are something altogether different. Solar flares don't emit much visible light, so they usually aren't detected in the visible part of the spectrum. Like sunspots, prominences often are associated with local strong magnetic fields.

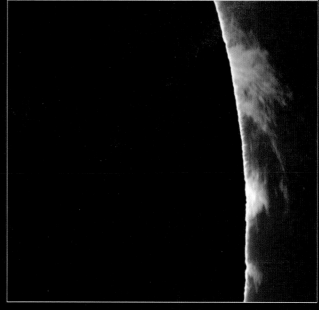

▲ *Solar filaments* (Glen Fountain)
▶ *Solar prominences* (Glen Fountain)
◀ *The sun in H-alpha light* (Jim Bonser)

How Old Is the Sun?

The sun's luminosity is 3.8×10^{26} watts. That means that each second, the sun radiates and hence must produce 3.8×10^{26} joules of energy. Where does that energy come from? The sun is powered by nuclear fusion. Deep in the sun's core, hydrogen fuses into helium, releasing energy. That comes at the expense of mass, following Albert Einstein's famous $E = mc^2$ equation. Each second, more than four billion kilograms (two million tons) of mass is converted to energy. Not to worry, the sun has 2×10^{30} kilograms of mass. However, there is a limit to how much hydrogen fuel there is in the sun's core — nearly ten billion years' worth. So, this would seem to fit the evolutionists' claim that the age of the earth, sun, and rest of the solar system is about 4.5 billion years.

FOR THUS SAYS THE LORD, WHO CREATED THE HEAVENS (HE IS GOD!),
WHO FORMED THE EARTH AND MADE IT (HE ESTABLISHED IT; HE DID NOT CREATE IT EMPTY,
HE FORMED IT TO BE INHABITED!): "I AM THE LORD, AND THERE IS NO OTHER."

Isaiah 45:18

However, there is a problem. As the sun fuses hydrogen into helium, the structure of the sun's core changes. This ought to increase the sun's output slowly, making the sun slightly brighter. Over thousands of years, this isn't significant. But over billions of years, this is important. This requires that the sun be 40% brighter now than when the sun was in its infancy. Most secular scientists think that life first emerged on earth about 3.5 billion years ago. But the sun would be 25% brighter now than back then. This would make the earth 17° Celsius warmer today than when life supposedly began 3.5 billion years ago. Since the earth's average temperature today is 15° Celsius, then 3.5 billion years ago the earth's average temperature would have been -2° Celsius, two degrees below freezing. If the earth's average temperature ever had been this cold, it would have been an ice ball. No one believes that. In fact, most scientists think that the earth's temperature has remained nearly constant all this time, with only minor variations. There is no satisfactory answer for what scientists call the faint young sun paradox. However, if the sun is only thousands of years old as the Bible indicates, then there is no paradox.

You may wonder why God made the sun with such a long-term source of energy, if the sun and everything else is only thousands of years old. The answer to that question likely lies in stability. By relying upon such a long-term energy source, the sun is very stable. If the sun used some other source of energy, it probably would not be stable. An unstable sun obviously would be a problem for living things on earth. Therefore, the sun's nuclear energy source appears to be part of God's design for earth.

◀ (Shutterstock)

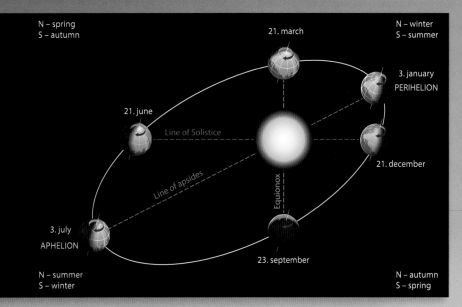

N – spring
S – autumn

21. march

N – winter
S – summer

3. january
PERIHELION

21. june

Line of Solistice

Line of apsides

Equionox

21. december

3. july

APHELION

23. september

N – summer
S – winter

N – autumn
S – spring

The earth's orbit around the sun is not a circle. Rather, the earth's orbit is an ellipse, with the sun at one focus. Consequently, the distance to the sun changes throughout the year. We are 3 % closer to the sun at perihelion (closest approach to the sun) than we are at aphelion (most distant point from the sun). Therefore, the apparent size of the sun ought to change throughout the year. This comparison of photos of the sun taken in H-alpha shows the difference in the apparent size of the sun between perihelion and aphelion. Perihelion occurs in early January, while aphelion is in early July.

Notice that the earth is closest to the sun during Northern Hemisphere winter and farthest from the sun when it is summer in the Northern Hemisphere. This slightly reduces the temperature difference between summer and winter in the Northern Hemisphere, thus moderating Northern Hemisphere temperature extremes. But what of the Southern Hemisphere, which is closest to the sun in summer and farthest from the sun during winter? That ought to make the difference in temperature between summer and winter greater than in the Northern Hemisphere. However, that is not the case — the temperature difference between summer and winter is about the same in either hemisphere. How can this be? A look at a globe reveals a fundamental difference between the two hemispheres. The Northern Hemisphere has a large amount of land mass, but the Southern Hemisphere is mostly water. Water has a high specific heat, which means that it takes a long time for water to heat and to cool. The large amount of water in the Southern Hemisphere moderates the extreme that would be present otherwise. This is an ideal situation, which strongly suggests design.

The size of the sun compared between perihelion and aphelion (Jim Bonser) ▶

Perihelion 01/04/2019

Aphelion 07/04/2019

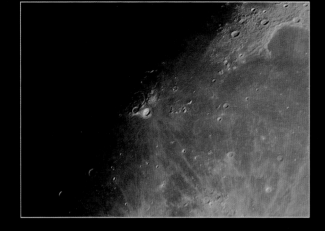

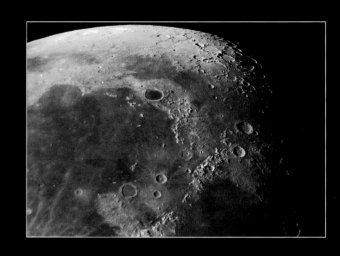

At only 240,000 miles away, the moon is our closest astronomical neighbor. The Day Four creation account of Genesis 1:14–19 gives several purposes for the heavenly bodies: to provide light on the earth; to separate night from day; and to be for signs, seasons, and days. The proper observance of time always has been an important function of astronomy. For instance, the British Royal Observatory at Greenwich and the U.S. Naval Observatory in Washington, D.C., were established for precise definition of time as an aid to navigation. So, what are some of the things that Scripture tells us about the moon?

the Moon

Mare Imbrium (Jim Bonser) ▲
Aristarchus (Jim Bonser) ◀
Purbach Cross (Jim Bonser) ▶

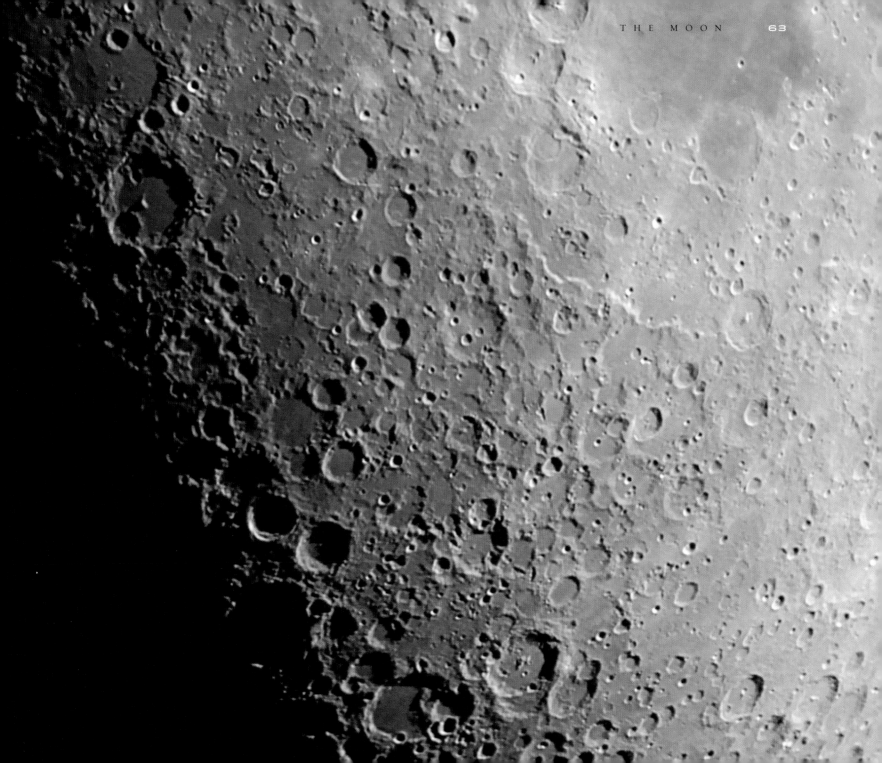

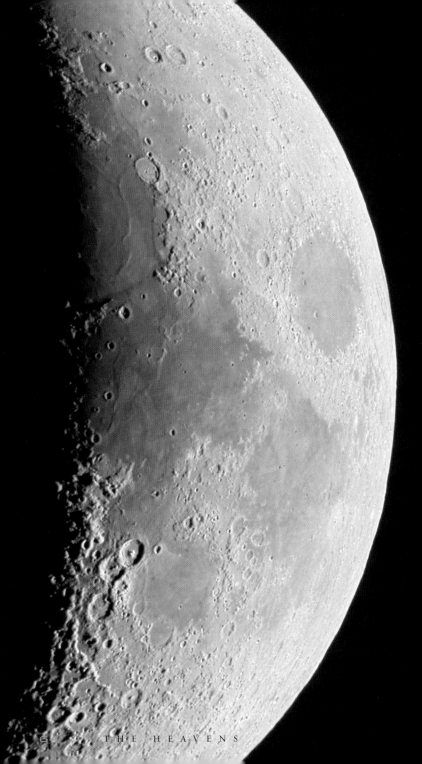

As the sun rules the day, the moon rules the night. But unlike the sun during the day, the moon is up at night only half the time.

... THE MOON AND STARS TO RULE OVER THE NIGHT,

FOR HIS STEADFAST LOVE ENDURES FOREVER. . . .

Psalm 136:9

Perhaps that is why God ordained that the stars rule the night along with the moon.

HE MADE THE MOON TO MARK THE SEASONS. . . .

Psalm 104:19a

Waxing crescent (Jim Bonser) ▲
Crescent (Glen Flountain) ▼
Waxing crescent (Jim Bonser) ◄
Gibbous (Glen Flountain) ▶

As mentioned above, one of the more important aspects of the moon is that it be for measurement of time. Today we usually associate the word "season" with the climatic seasons — spring, summer, autumn, and winter. However, seasons can refer to periods of time appointed for some purpose, such as baseball season or hunting season. This is the meaning of the Hebrew word translated "seasons" because it is the same word used for the festivals, such as Passover, instituted under the Law in the Old Testament.

ON THE THIRD NEW MOON AFTER THE PEOPLE OF
ISRAEL HAD GONE OUT OF THE LAND OF EGYPT, ON
THAT DAY THEY CAME INTO THE WILDERNESS OF
SINAI.

Exodus 19:1

The ancient Hebrews used a lunar or, more likely, a lunisolar calendar (the Jewish calendar today is nearly the same calendar). The month began when the thin crescent moon was first visible in the evening sky. This was the new moon that required a sacrifice.

... BESIDES THE BURNT OFFERING OF THE NEW MOON,
AND ITS GRAIN OFFERING, AND THE REGULAR BURNT
OFFERING AND ITS GRAIN OFFERING, AND THEIR
DRINK OFFERING, ACCORDING TO THE RULE FOR
THEM, FOR A PLEASING AROMA, A FOOD OFFERING TO
THE LORD.

Numbers 29:6

Waxing crescent (Jim Bonser) ▲
Waning gibbous (Jim Bonser) ▶
First quarter, far right (Jim Bonser) ▶

Since the moon's synodic period (the length of time over which the moon's phases repeat) averages 29.5 days, on the ancient Hebrew calendar each succeeding month varied between 29 and 30 days. On this calendar, full moon always falls on the 15th day of the month. According to Exodus 12, the first Passover was on the 15th day of the month of *Abib*, and God instituted *Abib* as the first month of the ceremonial calendar. In God's wisdom and timing, the Hebrews departed Egypt at night during full moon. The full moon undoubtedly provided light for them to see better as they traveled. The proper time to observe the other two major Hebrew feasts, Pentecost and Sukkot, similarly were timed to specific dates on the lunar calendar. Therefore, this function of the moon to be for seasons probably refers to the moon being used for the calendar.

More than 2,000 years ago, Julius Caesar abandoned the lunisolar calendar in favor of a solar calendar. On the Julian calendar (later modified to the current Gregorian calendar), the months were no longer tied to the moon's phases. Consequently, the Hebrew feasts jump around on our Gregorian calendar, but the feasts occur the same time each year on the Jewish calendar.

The moon is mentioned by name more than 60 times in the Bible. However, a little more than a third of those verses are about the new moon sacrifice (Deuteronomy 33:14), meaning the first of the month. Therefore, these verses do not refer to the moon per se. Nearly a third of the time that the moon is mentioned in Scripture, it is in prophetic passages describing God's judgment and possibly referring to future events (this probably falls under the purpose of being for signs). Hence, these verses don't tell us anything about the moon as it now exists. Four times the moon is described as an object of pagan worship, and the first time the moon is mentioned by name (Genesis 37:9), it is in a dream. About a quarter of the time that the moon is mentioned in the Bible, it is a direct reference to the moon, but many of those are in poetic passages that draw comparisons to some aspect of the moon. Consequently, the Bible doesn't directly reveal much about the moon as it exists. Hence, we are free to study the moon and infer much about its properties.

For instance, the most familiar feature on the moon is its many craters. Where did these craters come from? The best answer is that the craters formed from objects striking the lunar surface. But what were these bodies, where did they come from, and when did they fall onto the moon? Secular scientists who are committed to naturalism and billions of years have their answers to these questions. They think that the solar system formed from a cloud of dust and gas. Most of the gas and dust formed the sun, and the earth, moon, and other objects in the solar system formed from the remaining material.

Moon close-up (Jim Bonser)
Moon close-up (Jim Bonser)

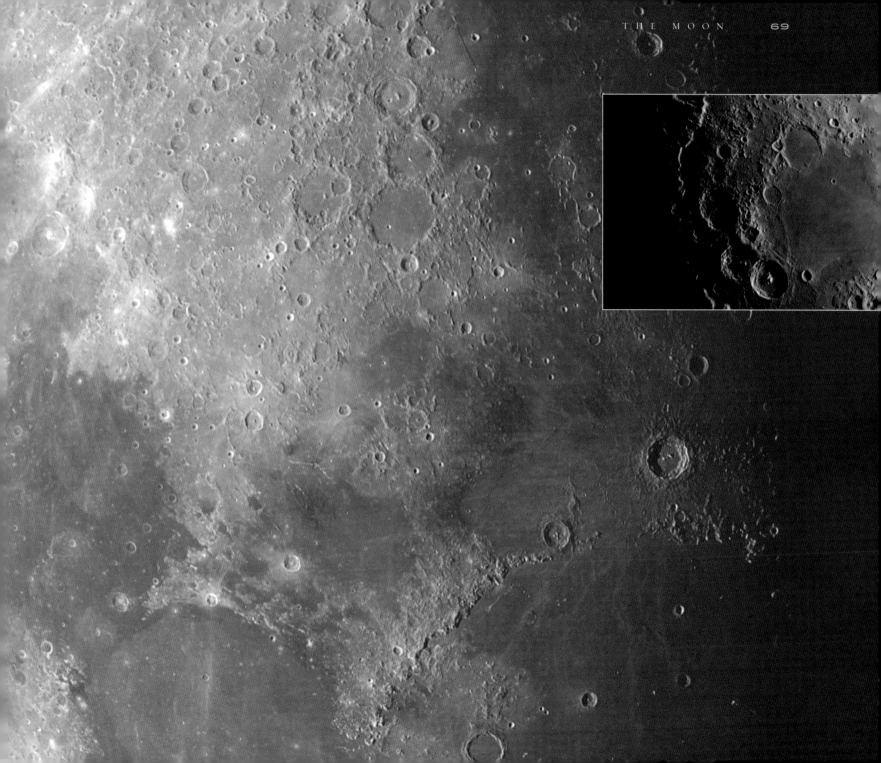

This process supposedly took much time, and not all matter from the original cloud formed into the bodies of the solar system right away. It was this remaining material that rained down on the moon and other bodies in the solar system that formed craters over the past 4.5 billion years.

However, those who are committed to the authority of Scripture see problems with this. For instance, we know that the world is only thousands of years old, not billions. Furthermore, God specially made the earth (Isaiah 45:18), and He made the earth three days before the sun, moon, and the rest of the solar system. So, impacts to form craters were not spread out over billions of years. Creation scientists have at least two views of when the moon's craters formed. Some craters may be from when God made the moon on Day Four, while other craters may have formed from later impacts during some catastrophe, such as the Flood of Genesis 6–9.

Another important lunar surface feature is its maria (pronounced ma'-ree-uh), which show up best at full moon. The maria are the darker, smooth regions on the moon. The heavily cratered lighter regions are the lunar highlands, so called because of their higher elevation above the maria. The maria and highlands give the "man in the moon" appearance of the full moon. The difference in color is due to difference in composition — the highland rocks resemble granite, while the maria rocks are like basalt. But why are there so few craters on the maria? It seems unlikely that the impacts missed the maria while striking the highlands.

Full moon (Danny Faulkner) ◄

Lunar libration (Jim Bonser and Glen Fountain) ▶

Notice that many of the maria are round, resembling large craters. The large maria that aren't so round appear to be the overlap of several round mare (pronounced ma'-ray, the singular form of maria). This suggests that late, great impacts formed the largest craters on the moon. These events were so catastrophic that they allowed molten material from deep inside the moon to rise and fill those large craters. Many creation scientists think that this happened at the time of the Flood. Therefore, creation scientists have a theory that can explain many features of the moon in terms of recent origin and young age.

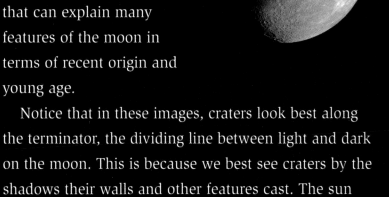

Notice that in these images, craters look best along the terminator, the dividing line between light and dark on the moon. This is because we best see craters by the shadows their walls and other features cast. The sun is either rising or setting along the terminator, so the shadows there are longest. Consequently, the worst time to look for lunar craters is at full moon because they cast no shadows that we can see then. The terminator usually is curved. This is because the moon is a sphere. This is the best evidence that the moon is a globe.

The moon rotates and revolves at the same rate. Consequently, one side of the moon always faces the earth. However, slight irregularities, called librations, allow us to peek around the moon's limb (edge) to see a little more than half the moon (a total of about 58%). These photos illustrate libration in longitude. While the moon rotates uniformly on its axis, due to its elliptical orbit, the moon doesn't revolve around the earth at a uniform rate. This allows us periodically to see a little beyond the east and west limbs of the moon. These two photographs are of the same phase, but they were taken more than three years apart. The arrows point to Mare Crisium. In the photograph on the left you can see farther beyond Mare Crisium than you can in the photo on the right.

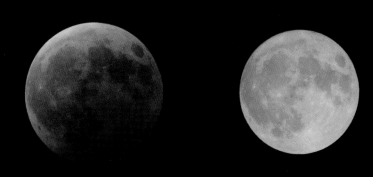

Just as the earth follows an elliptical orbit around the sun, the moon follows an elliptical orbit around the earth. The moon's orbit is more elliptical than the earth's orbit, so the moon's distance from earth varies by about 13%. These two photos show how much the apparent size of the moon changes between closest and most distant approach. The two photographs were taken during full moons seven months apart. The photo on the left was taken when the moon was near perigee, the point of closest approach of the moon to the earth. It also was during a total lunar eclipse, which is why the moon is so red. The photograph on the right was taken when the moon was near apogee, the most distant point from earth on the moon's orbit.

Moon perigee apogee (Danny Faulkner) ▲

TIDAL EVOLUTION

Most people are aware that the moon raises tides on the earth. But most people don't know that at the same time, the earth raises tides on the moon. Even fewer people know that there is an interaction of these tidal effects that leads to an interesting conclusion about the age of the earth and the moon. Tides on the earth act as a brake that gradually slows the earth's rotation. At the same time, the moon is accelerated in its orbit, causing it to slowly spiral away from the earth. These effects are feeble. The day is increasing at the rate of 0.0016 seconds per century. And the moon's distance is increasing about 4 centimeters per year. Over the course of thousands of years, this doesn't amount to much.

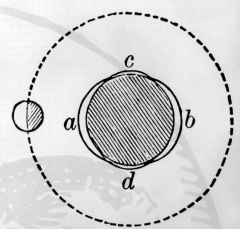

However, the tidal interaction is a steep function of distance. This means that in the distant past (many millions of years ago), the rate of change would have been much greater. The result is that this places a severe upper limit upon the earth and moon at not much more than a billion years. While this doesn't prove that the earth and moon are only thousands of years old, it indicates that they can't be billions of years old.

▲ *Lunar tide, also known as moon tide, is caused by the seas'*
gravitational forces.

The only two planets that can pass between us and the sun are Mercury and Venus. When these planets do this, we call it a transit. Because of its small size, when Mercury transits the sun, it appears very small. However, since Venus is much larger and closer to us, it appears much larger against the face of the sun.

were used to establish the scale of the solar system. The transit of Venus in 1639, as well as the pair of transits in 1761 and 1769 and the pair of transits in 1874 and 1882, were very important in this endeavor.

OTHER OBJECTS IN THE
Solar System

Transits of Mercury occur 13–14 times per century, but transits of Venus are much rarer. Transits of Venus follow a 243-year pattern. There are two transits separated by eight years, with alternating gaps of 105.5 years and 121.5 years. Historically, these rare events

However, advancements in technology have allowed astronomers to determine the scale of the solar system in better ways. Therefore, the recent pair of transits of Venus in 2004 and 2012 were more of a curiosity. If you missed these transits, you'll have to wait until the next pair, in 2117 and 2125.

The June 5, 2012 transit of Venus. Venus is the large dark spot to the upper left. (Jim Bonser) ▶

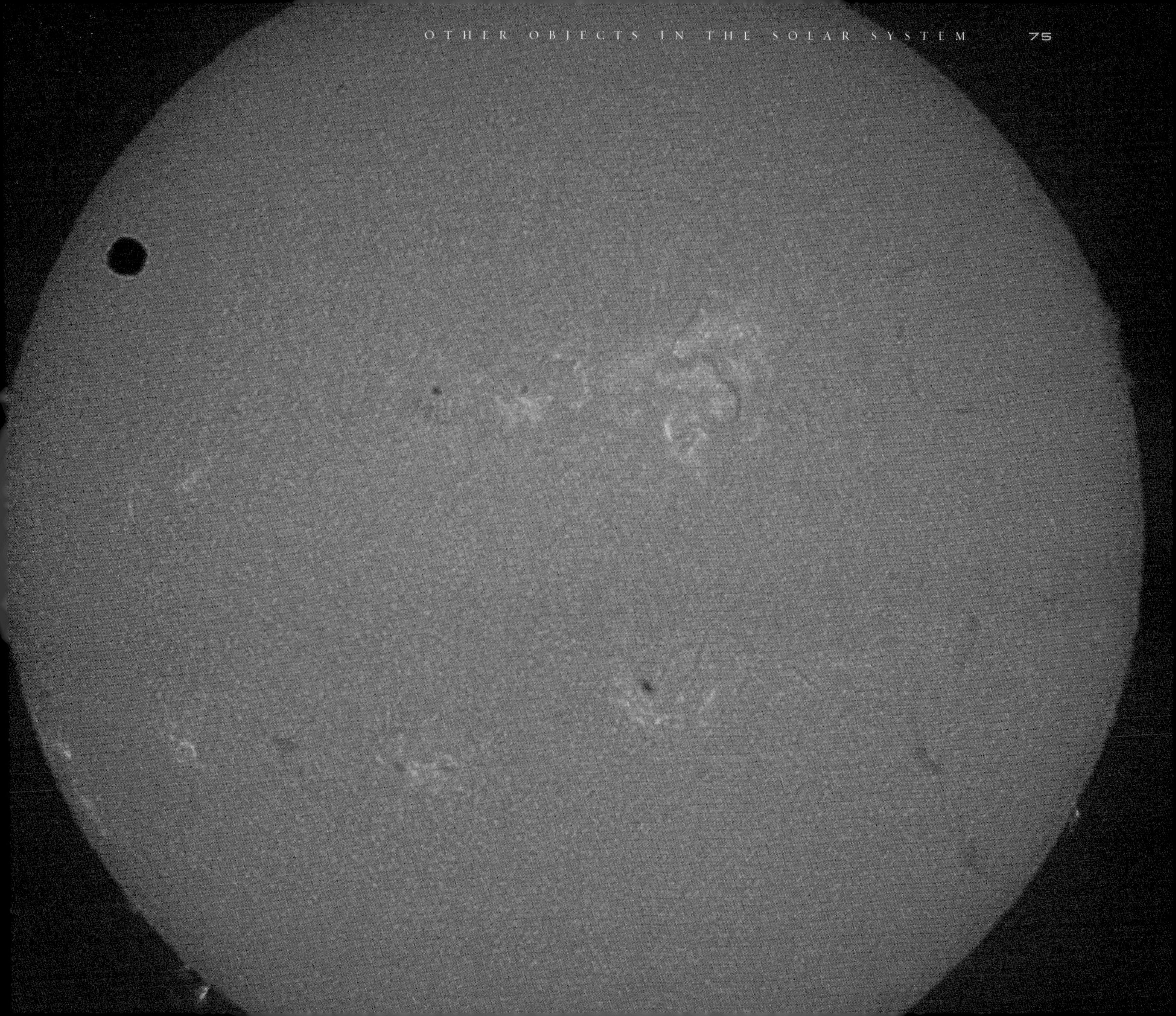

It is no accident that transits of Venus occur at such regular intervals. It is a complex interplay between the orbits of Venus and the earth. Part of this pattern is that as the earth orbits the sun eight times in eight years, Venus orbits the sun almost exactly 13 times. Astronomers say that there is an 8:13 orbital resonance between the two. There also is the matter of the slightly different orbital planes of the two planets. This makes transits possible each year only during two brief intervals six months apart. And then only if Venus passes between the earth and sun during this time (an event astronomers call an inferior conjunction). Some people may say that this sort of pattern must be present in the world, so it signifies nothing. However, this assumes that the world *must* be governed by regular patterns, such as gravity. But why must the world be this way, as opposed to being chaotic? The naturalist has no answer for this question, other than the fact that it's just the way that the world is. However, Scripture makes it clear that God designed the world, and hence there is order. Apart from creation, this order makes no sense.

Jupiter (Jim Bonser) ▲
Jupiter (Glen Fountain) ▼

look at with a telescope. Even a small telescope will reveal its four largest natural satellites, or moons. They look like four small stars along a line passing through Jupiter. As they orbit Jupiter, their positions change each night. Sometimes there are two on one side of Jupiter, with two on the other side. Sometimes it all four satellites are lined up on one side of Jupiter. And then sometimes, there are only three satellites visible. Where is the other one? It is either behind the planet or in front of it. But if you wait a while, it might emerge and become visible again.

Sometimes one can see the shadow of one of Jupiter's satellites, or moons, as it crosses the planet. This is a solar eclipse on Jupiter. Too bad there isn't anyone there to watch it.

With a sufficiently large telescope, dark bands in Jupiter's atmosphere may be visible. These bands are aligned with Jupiter's equator and line up with Jupiter's satellites. The bands are a consequence of Jupiter's rapid rotation — a day on Jupiter is less than ten hours. An even larger telescope may show Jupiter's famous red spot, shown here to the left. However, since our color-sensitive cones require much light, people rarely see any color in the red spot.

▲ *Jupiter with two of its natural satellites, or moons, top* (Glen Fountain)
▲ *Jupiter with the shadow of one of its natural satellites, bottom* (Glen Fountain)

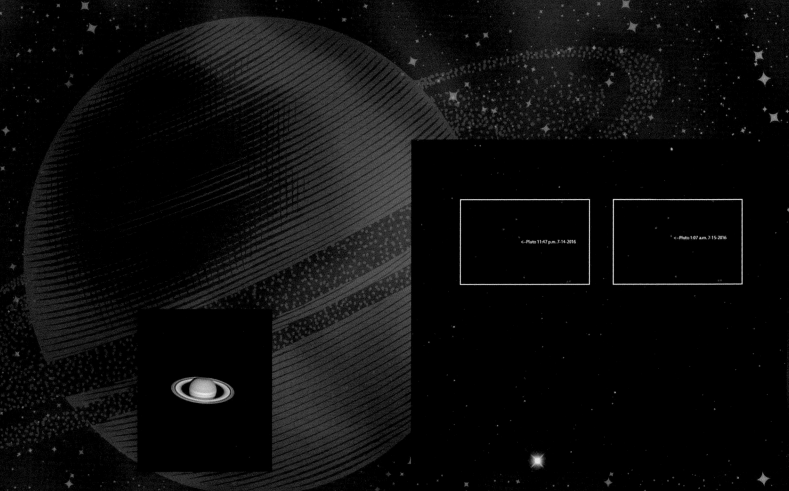

<--Pluto 11:47 p.m. 7-14-2016

<--Pluto 1:07 a.m. 7-15-2016

My absolute favorite thing to see through the telescope is Saturn. Even a small telescope will reveal Saturn's rings, though you will need magnification of at least 30x to see it. I first saw Saturn's rings a half century ago, and though I've looked at them thousands of times since, I never get tired of looking at Saturn.

Pluto is a very faint object. I've seen it only a few times, and then with a very large telescope. Even with the largest telescopes, Pluto appears merely as a tiny speck. Here we see the change in Pluto's position over less than two hours during one night in July 2016.

Pluto (Jim Bonser) ▲
Saturn (Glen Fountain) ◄

For much of my life, Pluto was a planet. However, in 2006, astronomers officially decided that Pluto wasn't a planet. Why? From its discovery in 1930, astronomers realized that there were problems with Pluto. The main difficulty is that Pluto is just too small to be a planet — it would take 22 Plutos to match the mass of Mercury, the smallest planet.

If Pluto is not a planet, then what is it? There are many small bodies in the solar system. In fact, astronomers are now aware of nearly a million SSSBs (Small Solar System Bodies), as these objects are officially called. Most of them are asteroids, while others are comets. The differences between these two groups are composition and their orbits, though there can be some overlap. Asteroids have rocky composition and orbit the sun in nearly circular orbits that are close to the same plane as the planets orbit the sun. The characteristics of Pluto most closely match those of asteroids.

Comets have an icy composition. Their orbits often are highly inclined to the plane of the solar system. And their orbits are very elliptical, taking them alternately very far from the sun and very close to the sun. Consider Comet Lovejoy, seen here with the Dolphin Nebula to the upper left. In the constellation Cassiopeia, the Dolphin Nebula is a planetary nebula with an unusual shape. Comet Lovejoy was visible in late 2013 and early 2014.

Comet Lovejoy (Jim Bonser) ▲
Lovejoy comet with Pleiades (Shutterstock) ▶▶◀

A comet consists of a small nucleus made of various ices with much dust mixed in. Since a comet normally is far from the sun, it is very cold and hence remains frozen most of the time. However, once each orbit, the nucleus comes close to the sun. The sun's radiation heats the ices, causing them to sublime (going directly from solid to gas). The gas expands, and the sun's ultraviolet light ionizes the gas. As electrons recombine with atoms in the gas, they give off visible light. This is the bright coma visible in the photograph. The tiny comet nucleus is deep inside the coma. The sun's radiation and solar wind push gas and dust particles away from the sun to produce the tails of the comet. As you may infer, the sun must be to the lower left in this photograph.

The orbital period of Comet Lovejoy is about 7,000 years. How many times could it orbit the sun? Since the comet loses much of its matter each time it passes close to the sun, it could not orbit very long. In the case of Comet Lovejoy, it couldn't have been orbiting for anywhere near a million years, let alone the supposed 4.55-billion-year age of the solar system. To solve this dilemma, evolutionary astronomers think that Comet Lovejoy and many other comets came from the Oort cloud, a distribution of many comet nuclei orbiting very far from the sun. However, there is no evidence that the Oort cloud exists. This problem for the supposed old age of the solar system isn't a problem if the solar system is young. Therefore, the existence of comets such as Comet Lovejoy is evidence that the solar system is young.

Comet Hale-Bopp was much expected in 1997.
It was one of the brightest comets in a very long
time, and many people around the world saw it.
But as the world awaited the arrival of Comet
Hale-Bopp, Comet Hyakataki suddenly appeared
in 1996. It was unusual to have two bright
comets appear so close together in time.

Visible in 2020, Comet NEOWISE was the
brightest comet in recent years. Unfortunately,
this comet was very low in the morning and
evening sky, making it difficult to see.

▲   *Comet NEOWISE* (Danny Faulkner)
◀   *Comet Hale-Bopp* (Jim Bonser)
▶   *Comet Hyakataki.* (Deb Bonser)

There are many microscopic particles of dust in the plane of the solar system. The particles likely are pieces broken off of asteroids and comets. The dust scatters light from the sun, producing a faint glow along the plane of the solar system. This part of the sky is called the zodiac, so we call this light the zodiacal [zo-DIE-uh-cuhl] light. The zodiacal light is easiest to see before dawn in the east near the autumnal equinox or after dusk in the west near the vernal equinox. It is important to have very clear, dark skies with no moon and a good exposure down to the horizon. The zodiacal light can be brighter than the Milky Way, though it is more difficult to see than the Milky Way. There are at least two reasons for this. One reason is that the zodiacal light is more diffused with no definite edge and no dark lanes like the Milky Way has. This lack of contrast makes the zodiacal light easy to miss. Second, the brightest part of the zodiacal light is close to the horizon. When objects are so low in the sky, the earth's atmosphere absorbs much of their light, making them appear dimmer (for instance, the sun appears dimmer near sunrise and sunset than when high in the sky).

I took this photograph on the morning of September 1, 2019, from 8,300 feet elevation in Arizona. The high elevation and dry weather minimized the amount of dimming by the earth's atmosphere. The zodiacal light is the triangular region of light silhouetting the trees on the left and rising to the upper right. The Praesepe, or Beehive star cluster (M44, NGC 2632) is the fuzzy spot to the right of the tallest tree. Procyon is the bright star to the right of the tallest tree. Castor and Pollux are the two bright stars to the upper right of the tallest tree.

Solar radiation has a profound effect on dust particles. Solar radiation pushes the tiniest particles outward, away from the sun. This is the force that produces the dust tails of comets. For larger particles, up to about a millimeter in diameter, solar radiation produces a retarding force called the Poynting-Robertson effect. This retarding force causes dust particles to spiral inward toward the sun. These two effects act to cleanse the solar system of dust particles. Depending upon the rate with which new particles are introduced, the existence of dust in the solar system may be evidence that the solar system is far younger than billions of years.

The zodiacal light (Jim Bonser) ▶

M eteors sometimes are called "falling stars" or "shooting stars." They certainly look like stars, and until a few centuries ago, they would have been classified as stars. Only in modern times have we decided that meteors are something quite different from stars. Meteors are caused by small, fast-moving particles of rock or ice from space that burn up in earth's upper atmosphere. In this sense, they represent the intersection of earth with space. So, the word "meteor," which comes from a Greek word related to the atmosphere, is aptly chosen.

WHERE Space AND Earth CONNECT

Many meteors occur randomly, but there are periods of time each year when an unusual number of meteors happen. We call this a meteor shower. If we trace the tracks of meteors backward, they appear to diverge from a point in space called the radiant of the shower. The location of the radiant lends its name to a meteor shower. For instance, one of the most reliable meteor showers is the Perseid shower each August. It is named for its radiant, located in the constellation Perseus. The Perseids, as the meteors of this shower are called, are from debris in the orbit of comet Swift-Tuttle. Each August, the earth crosses the orbit of this comet, producing many splendid meteors.

This bright meteor (left) was from the Perseid shower. It happened at 1:47 a.m. EDT on Friday, August 14, 2015. The meteor traveled upward and to the right from the radiant, which is to the lower left. Notice that the color and brightness of the meteor varied over its path. This is very common. An artificial satellite left a faint streak to the left of the meteor's trail during this 30-second exposure.

The meteor below also is a Perseid meteor. This 10-second exposure was taken shortly after 3:00 a.m. MST on Monday, August 13, 2018. The Andromeda galaxy can be seen at the center of this 10-second photograph.

◄ *Meteor* (Danny Faulkner)
► *Meteor* (Danny Faulkner)

This is another Perseid meteor passing through the constellation Cygnus.

All the host of heaven shall rot away,
and the skies roll up like a scroll.
All their host shall fall,
as leaves fall from the vine,
like leaves falling from the fig tree.

Isaiah 34:4

IMMEDIATELY AFTER THE TRIBULATION OF THOSE DAYS THE SUN WILL BE DARKENED, AND THE MOON WILL NOT GIVE ITS LIGHT, AND THE STARS WILL FALL FROM HEAVEN, AND THE POWERS OF THE HEAVENS WILL BE SHAKEN.

Matthew 24:29

... AND THE STARS OF THE SKY FELL TO THE EARTH AS THE FIG TREE SHEDS ITS WINTER FRUIT WHEN SHAKEN BY A GALE.

Revelation 6:13

Do these apocalyptic passages refer to an unusually intense meteor shower? Many Bible scholars think that they may. This would be consistent with the stars being for signs (Genesis 1:14).

Meteor (Danny Faulkner)

Another place where the space and earth intersect is aurorae, or what are sometimes called the "northern lights," at least in the Northern Hemisphere (in the Southern Hemisphere the term is "southern lights"). Aurorae are caused by fast-moving charged particles from the sun. The solar wind is an outrush of these particles from the sun, taking a day or two to span the distance between the sun and earth. How these fast-moving particles are accelerated is not entirely clear. The earth's magnetic field normally deflects the paths of these particles to protect the earth from their harmful effects. Some of the particles follow magnetic field lines to where they come closest to the earth at high latitudes. About 60 miles up, some of these particles ionize atoms of gas in the earth's atmosphere. The recombination of electrons with the ions releases energy in the form of visible light. This is an aurora. Occasionally, violent eruptions on the sun release far more charged particles than normal. If these eruptions are directed toward the earth, a magnetic storm ensues, and aurora activity can be far higher than usual, and aurorae can be seen much farther south than usual.

◄ *Aurora* (Jim Bonser)

▲ *Aurora* (Jim Bonser)

Lunar Eclipses

Sometimes the earth passes between the sun and moon so that the earth's shadow falls on the moon. This is a lunar eclipse. As the earth's umbra (shadow) creeps across the lunar surface, it is very clear that the earth's umbra is round. The only shape that always casts a circular shadow is a sphere. Therefore, the earth must be a sphere. This is not a modern argument for the earth's globe shape because Aristotle wrote about this nearly 2,400 years ago.

Finally, the moon is entirely immersed in the earth's umbra — a total lunar eclipse. But a totally eclipsed moon rarely is completely dark. Instead, a totally eclipsed moon usually is some shade of red. Sometimes it is orange like a pumpkin, while other times it looks like copper. This is because the earth's umbra is not entirely dark. The earth's atmosphere bends light around the earth and into its shadow. Every lunar eclipse is different. For instance, the two single photos of totality here, one from September 2015 and the other from January 2019, are different colors. Furthermore, one is light on the top, while the other is light on the bottom. This is because in one eclipse (2015) the moon passed through the lower portion of the earth's umbra, while in the other eclipse (2019) the moon passed through the upper portion of the earth's umbra.

▲   *The total lunar eclipse of September 27, 2015* (Deb Bonser)

◀   *Lunar eclipse* (Answers in Genesis staff)

THE SUN SHALL BE TURNED TO DARKNESS, AND
THE MOON TO BLOOD, BEFORE THE GREAT AND
AWESOME DAY OF THE LORD COMES.

Joel 2:31

. . . THE SUN SHALL BE TURNED TO DARKNESS AND
THE MOON TO BLOOD, BEFORE THE DAY OF THE LORD
COMES, THE GREAT AND MAGNIFICENT DAY.

Acts 2:20

WHEN HE OPENED THE SIXTH SEAL, I LOOKED,
AND BEHOLD, THERE WAS A GREAT EARTHQUAKE,
AND THE SUN BECAME BLACK AS SACKCLOTH, THE
FULL MOON BECAME LIKE BLOOD. . . .

Revelation 6:12

Since blood is red, and some shade of red is the most common color of a total lunar eclipse, some Christians think that these three apocalyptic verses refer to a total lunar eclipse. However, not all lunar eclipses are red, and even when red, the color isn't that suggestive of blood. Furthermore, the original readers of these passages were very familiar with the sacrifices at the Temple. There was no clean-up between sacrifices at the Temple, so the place of sacrifices became stained with dried blood. Dried blood is very dark, almost black.

THE EARTH QUAKES BEFORE THEM;
THE HEAVENS TREMBLE.
THE SUN AND THE MOON ARE DARKENED,
AND THE STARS WITHDRAW THEIR SHINING.

Joel 2:10

Several biblical passages speak of the moon being darkened, as well as the sun and the stars. Perhaps the three verses that speak of the moon being turned to blood refer to this event as well. Could this be a lunar eclipse? Not likely. Such prophetic passages suggest something unexpected and inexplicable. But lunar eclipses are well understood, and we can predict when and where they will be visible many years in advance. Hence, these prophetic passages probably are talking about something entirely different, something that we won't anticipate, nor can we explain.

The total lunar eclipse of January 20, 2019 (Jim Bonser) ▶

The moon must be full for a lunar eclipse, but an
eclipse doesn't happen every full moon. The reason is
that the moon's orbit around the earth is tilted about
5 degrees to the earth's orbit around the sun. During
most full moons, the moon is too high or too low for
the earth's umbra to eclipse it. Only if the moon is
near one of its two nodes, the places that the
moon's orbit crosses the earth's orbit, is
an eclipse possible. Twice a year the
direction where full moons can occur
are up with the moon's
nodes.

We call these times eclipse
seasons. When a full moon happens
during an eclipse season, then some
sort of lunar eclipse visible somewhere on
the earth results. Eclipse seasons are six months apart,
and they last a little more than a month each.

On August 21, 2017, millions of people crowded into a swath of land no more than 70 miles wide extending from Oregon to South Carolina. The event was the first total solar eclipse visible from the continental United States since 1979. It was the first coast-to-coast total solar eclipse in a century. A total solar eclipse is perhaps the most wonderful experience in all creation. Most people understand that it gets dark during totality, but there is far more than this to a total solar eclipse. The total phase only lasts a few minutes (about 2½ minutes at most for the August 21, 2017, eclipse). Totality is preceded by a little more than an hour of the partial phase as the moon slowly creeps across the sun's photosphere (surface). The photosphere is extremely bright, so it is not safe to view directly with our eyes. Most people choose to view the partial phases with specially designed eclipse filters. Only when the last bit of the photosphere is covered, and totality begins, is it safe to look toward the sun with no filters.

The partial phase of the eclipse with sunspots (Danny Faulkner) ▶

If the sky is very clear, then when the moon covers about 70% of the sun, some eerie effects are noticeable. The amount of sunlight is reduced as when thin clouds obscure the sun, or when the sun is low in the sky. However, those conditions normally produce indistinct shadows, but if anything, during a partial solar eclipse the shadows are more distinct than ever. Somewhere after 90% coverage, street lights may come on, and some animals may begin to behave as if evening is approaching. Even with the gradually fading light, no one is quite prepared for the sudden drop in light in a matter of seconds as totality sets in. Totality is not as dark as midnight. Rather, it is more like dusk, with the brighter stars and planets visible. Instead of orange and red on the western horizon at dusk, a 360-degree panorama of sunset often is seen during totality.

Totality showing the "sunset" on the horizon (Jebi Koilpillai) ◁
Totality, lower left (Deb Bonser) ▷
Totality, upper right (Jim Bonser) ◁

But all this wonder is dwarfed by the view of the sun itself. With the bright photosphere blocked, the much fainter light of the solar corona, the outermost layer of the sun, bursts into view. The corona is pearly white and extends out several solar diameters. The corona is a very thin gas, and hot — more than one million degrees Fahrenheit. Such high temperature highly ionizes the gas. With the presence of the sun's magnetic field, the corona is a plasma, and the interaction of fast-moving charges with the magnetic

field is the dominant force in the corona. Many streamers pass outward through the corona, mapping out magnetic field lines there. The corona changes rapidly over time, so its appearance is different at every eclipse. With increasing distance from the sun, the solar corona gradually thins and morphs into the solar wind, an outrush of fast-moving charged particles flowing away from the sun.

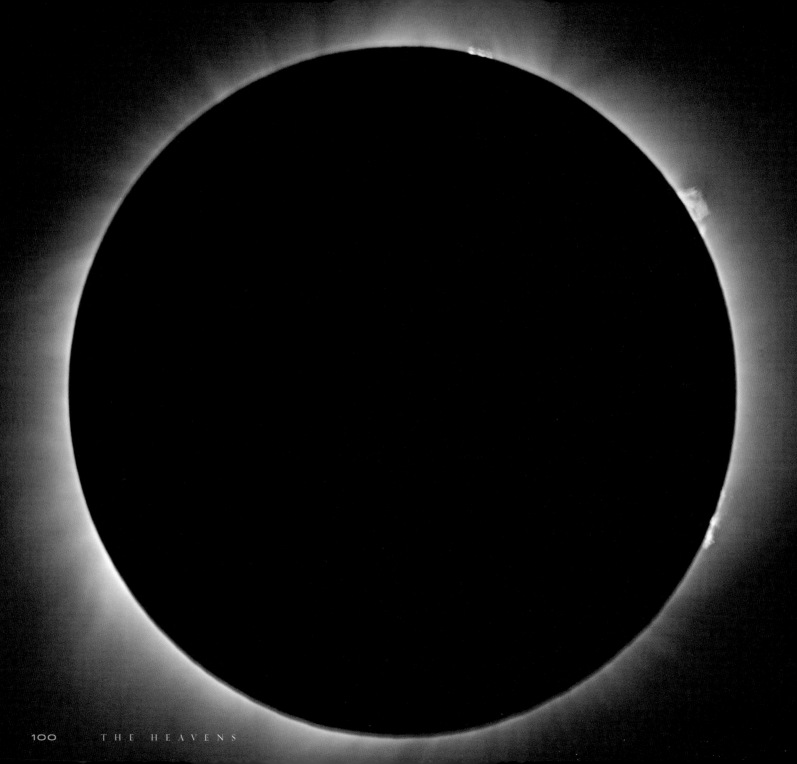

Much closer to the sun's limb (edge) are solar
prominences, deep red loops extending from
the photosphere out into space. Like the corona,
prominences are formed and sculpted by magnetic
fields on the sun. With special filters, it is
possible to view solar prominences
when there is no eclipse.
However, those views are
not nearly as complete or
breathtaking as during a
total solar eclipse. In similar
manner, with special
equipment it is possible for
astronomers to observe the
outer portions of the solar
corona any time. However,
we cannot see the inner corona
except during totality. The inner
corona is the most active and rapidly
changing part of the corona, so besides the
beauty and wonder of total solar eclipses, they offer
astronomers the only opportunity to study the most
dynamic part of the corona.

▲   *Totality* (Jim Bonser)
◄   *Totality* (Danny Faulkner)

For a fraction of a second at the beginning and ending of totality, Bailey's beads and the diamond ring effect are visible. Both are caused by a tiny bit of the photosphere peeking through irregularities, such as craters, on the moon's limb. Bailey's beads appear as a string of pearls along a small arc of the lunar limb.

The diamond ring is a burst of light that looks like a diamond, while the ring consists of the innermost part of the corona completely around the sun and moon. Both Bailey's beads and the diamond ring are very dangerous to look at, so no one ought to attempt to see them without proper filters.

Diamond ring (Deb Bonser) ▲
The diamond ring at the February 26, 1979 total solar eclipse (Danny Faulkner) ▶

Before one knows it, the all-too-brief totality ends, and the light of day begins to return. Totality is followed by partial phase, and soon people pack up and head home — except that with millions of people wanting to drive home at the same time, the roads were clogged for hours. Not to worry, all were in high spirits from the euphoria of experiencing God's handiwork. And handiwork it is. The sun's diameter is 400 times larger than the moon's diameter. But the sun is 400 times farther away than the moon is. Consequently, the sun and moon appear the same size in the sky — about ½ degree. Therefore, when the moon covers the sun, it just barely does so. If the moon were a little smaller or farther away, there would be no total solar eclipses.

But if the moon were larger or closer, solar eclipses would be grossly over-total. That is, the beautiful prominences and corona would not be visible, so total solar eclipses would not be nearly as spectacular. And total solar eclipses would be visible over much larger portions of the earth, making them more common. While a total solar eclipse occurs on average about every year and a half somewhere on the earth, the path of totality of each eclipse is very narrow, so very little of the earth may experience any given eclipse. For any location on earth, a total solar eclipse is seen about once in four centuries on average. Many of the other planets have natural satellites, or moons (there are about 200 known), but only on the earth are the conditions of rarity and extreme beauty combined. And only on the earth does it matter, because only on earth are there creatures to appreciate them. One can believe this merely is coincidence. However, I think total solar eclipses are a precious gift of God that can lead us to understand that the heavens do declare God's glory.

DANNY FAULKNER

Danny Faulkner received his Ph.D. in astronomy from Indiana University. He is distinguished professor emeritus at the University of South Carolina Lancaster (USCL), where he taught astronomy and physics for more than a quarter century. Upon his retirement from USCL at the end of 2012, Professor Faulkner became the staff astronomer at Answers in Genesis in northern Kentucky. He specializes in stellar astronomy, particularly the study of binary stars. Professor Faulkner has published more than a hundred papers in astronomy journals. He also is the author of numerous articles and a few books about creation as it relates to astronomy.

Professor Faulkner was fascinated with astronomy as a young child. This interest grew as he became an avid amateur astronomer by his high school years. Born into a Christian home, Professor Faulkner was born again at age six. During his sophomore year of high school, Professor Faulkner came to realize that his calling was to be an astronomer for God's glory. This decision set him on a path to pursue the required education to do this.

Professor Faulkner wrote the text of this book, and he supplied some of the photos. However, most of the photographs came from the work of others. Professor Faulkner is indebted to their contributions.

MEET THE Photographers

Interestingly, Professor Faulkner didn't take up astrophotography until a few years ago. He reckons that he saved a lot of time by waiting until digital single lens reflex cameras became available. He is amazed with how sensitive modern cameras are compared to the sensitivity of film. And he also is pleased with the nearly instant results, which allows a person to make corrections on the spot instead of waiting until much later to find out what went wrong. This often makes the difference between getting a good photograph or getting at best a lousy photograph of a rare event. During the few years that Professor Faulkner has been doing astrophotography, he has learned many of the tricks that have greatly improved his results. And he often is pleased with those photographs, at least until he sees the photographs of amateur astronomers. Those photographs often leave him with the question, "Why do I even bother?" You'll meet some of them next.

Danny Faulkner's wonder and amazement captured during the 2017 total solar eclipse (Jebi Koilpillai)

GLEN FOUNTAIN

Glen Fountain started his trek in astronomy when he was around 10 years old, after receiving a small red Tasco Refractor telescope. While this specific telescope did not exactly meet his expectations, it was a small stepping stone to the desire to see more. In 1998, Glen and his wife Katrina were married and bought their first "real" telescope, a Celestron C8 with clock-drive. They had no idea how to operate it, but as they spent many nights under the stars from their front yard of an old farmhouse in Toccoa, Georgia, they began to slowly learn and see some amazing sights. Without any formal guidance and very little knowledge, everything they saw at first looked like a possible new discovery. (They had no clue!) Several years later they purchased another telescope which had a computerized "GoTo" feature. Now all that was required was to let the handheld controller slew their telescope across the sky to some celestial object that was waiting to be seen. This greatly helped to speed up their learning curve and fostered a more intimate knowledge of the night sky. Armed with this powerful tool, they were now able to see many new and wonderful creations of God and learn about His night sky all at the same time. Due to this helpful technology, along with the reading of many books, Glen could now identify the Big Dipper if given a few hints and pointed in the right direction.

In 2007, Glen began his first attempts in astrophotography. His results were not very good, but they were a way to show others God's creation. After a couple years of attempting to take longer exposures so that faint objects could be seen, he discovered that his desire far outweighed his ability. However, in 2013, when their son Micah was born, Glen began exploring ways he could continue this hobby and easily share it with Micah as he grew older. It was the next year, in 2014, after his small observatory was built from some plans he purchased off the internet, that he and his family could easily view the heavens on any clear night whenever they wanted. (For Glen, this was pretty much every chance he got so he could practice taking pictures of the night sky.) He named his observatory "Ex Nihilo," which is from the Latin theological phrase, Creatio Ex Nihilo, meaning "creation out of nothing." This biblical concept in Hebrews 11:3 expresses that God did not use anything pre-existing, but rather He created everything by only His word.

Glen attended Toccoa Falls College and graduated with a degree in Youth Ministries. He has not received any formal education in the field of astronomy, except what he was taught in a basic high school science class. While he is extremely grateful for all the education he received, he is most thankful for the biblical foundation that he was given at home by his father and mother, Eugene and Pat Fountain. It was due to their godly example and sharing of the gospel that Glen repented of his sins and put his faith in Jesus on June 20, 1990, at the age of 14.

One final thought. Anyone who has read through this book and witnessed God's amazing handiwork can, with basic knowledge, observe many of these same objects from their very own backyard. The more one looks through a telescope, a pair of binoculars, or even with just the naked eye, the more they will learn and appreciate our Lord's incredible creation. And if there is a desire to share all of this with friends and family who refuse to stay up all night to see it, then go grab a camera and start the fun journey of astrophotography!

JIM BONSER

Jim Bonser received a B.A. in Bible from Central Bible College in Springfield, Missouri, and an A.A. in Business and an A.S. in Computer Science from Marshalltown Community College in Marshalltown, Iowa. He has served the Lord as pastor at Stavanger Friends Church in rural Marshall County, Iowa, since 2006.

Bonser has always been curious about the world around him. As a child, he collected rocks and butterflies and other insects and loved to pick up rocks to see what was hiding underneath. On his 9th birthday, he received a telescope as a gift. At first, he was disappointed that although the little scope showed him more stars than he could see with his naked eye, they all still looked like stars. However, that all changed when he decided to aim the little 50mm scope at a bright, soft, yellow-colored star. To his great surprise, it turned out not to be a star at all, but instead, it was the jewel of the solar system: Saturn! It was very small in the little scope, but it was unmistakable, and Bonser was hooked. His aunt, who was a schoolteacher, bought him a beginner astronomy field guide, which he devoured.

The pictures of the planets and star clusters and nebulae in the book inspired him to go searching for them with his little scope, but he soon became discouraged because other than Saturn, he could not find any of those beautiful things pictured in the book. Fast forward to 1990 when Bonser wanted to share the planets through a telescope with his elementary-aged kids. His childhood scope was long gone, but he located an amateur astronomer who not only helped show Saturn and Jupiter to the kids but ended up selling Bonser an 8-inch Schmidt-Cassegrain telescope, restarting Bonser's childhood love of the night sky. Bonser convinced his wife that although the telescope was a bit expensive, after that, the sky is free! She is still wondering when it will be free.

Bonser now had a good telescope capable of astrophotography and, even more important, a friend who knew his way around the sky. Ever since acquiring the scope (and the friend), Bonser has been improving his imaging techniques and equipment. He loves using the images as an opportunity to share the beauty and wonder of what God has created and placed in the heavens. While not Hubble Space Telescope quality, Bonser is amazed at how much better many of his images are than the professional images that inspired him in that book his aunt gave him so many years ago. Bonser considers himself blessed by God for being given the opportunity to image the work of God's fingers in space. Glory to God!